SYNAPSIDA

Also by John C. McLoughlin
THE ANIMALS AMONG US
ARCHOSAURIA

JOHN C. MCLOUGHLIN   A New Look into

SYNA

the Origins of Mammals   THE VIKING PRESS

PSIDA

Copyright © John C. McLoughlin, 1980
All rights reserved
First published in 1980 by The Viking Press
625 Madison Avenue, New York, N.Y. 10022
Published simultaneously in Canada by
Penguin Books Canada Limited

Grateful acknowledgment is made to Holt, Rinehart and Winston,
Publishers, for permission to reprint a selection from *The Green Pastures*
by Marc Connelly. Copyright 1929, 1930, © 1957, 1958 by
Marc Connelly.

LIBRARY OF CONGRESS CATALOGING IN PUBLICATION DATA
McLoughlin, John C.
Synapsida.
Bibliography: p.
Includes index.
1. Pelycosauria. 2. Therapsida. I. Title.
QE862.P3M24   567.9′3   79–56270
ISBN 0–670–68922–X

Printed in the United States of America
Set in Linotype Granjon

CONTENTS

ACKNOWLEDGMENTS vii

PREFACE viii

EVOLUTION AND ITS DISCOVERY 1

ROOTS 6

TRANSITION 13

COTYLOSAURIA 20

ENTER THE SYNAPSIDA 27

ANOMODONTIA 54

BRINGING HOME THE PERMIAN BACON 69

THE MESOZOIC 89

THE MAKING OF THE MAMMALS 100

GLOSSARY 135

BIBLIOGRAPHY 142

INDEX 143

This book is for A.W.D. and B.B.M.
and the school in which they and their colleagues
so elegantly teach the art of inquiry
and the science of seeing.

ACKNOWLEDGMENTS

In preparation of a book such as this, it is a special pleasure to acknowledge the assistance of the experts to whom we owe most of our appreciation of the creatures and processes discussed herein.

Dr. Nicholas Hotton III, research curator of the Department of Paleobiology at the Smithsonian Institution, provided me not only with many valuable references but with a lively correspondence throughout the preparation of this work. Dr. Farish A. Jenkins, Jr., and Dr. A. W. Crompton, both of the Museum of Comparative Zoology at Harvard University, offered a matrix of research materials embracing nearly the entire current body of knowledge concerning the reptile-mammal transition. In his book *Vertebrate Paleozoology*, Dr. Everett C. Olson of the Department of Biology of the University of California at Los Angeles furnished the basis for yet deeper examination of the story of synapsid evolution and its relationship to our own existence.

Mention must also be made of the untiring labors of Dr. Edwin H. Colbert, curator emeritus of vertebrate paleontology at the American Museum of Natural History and present curator of vertebrate paleontology at the Museum of Northern Arizona, Dr. James Hopson of the University of Chicago, the late Alfred Sherwood Romer of the Museum of Comparative Zoology at Harvard, and all the others who have contributed to our understanding of the strange times and places visited in this book. It is due largely to the splendid efforts of such as these, and to the personnel of the great institutions that support their work, that we approach the self-awareness necessary to the continuance of human and humane life on this planet.

On a more personal level, my thanks go to Amy Pershing, for knitting it all together; to John Lander Merrill and his family, for keeping me on a relatively even keel; to Dr. Philip L. Shultz, for seeing it into a comfortable landing; and to Carole and Ariana, for lending it purpose.

PREFACE

The human self-image has progressed from one of smug complacence at the center of creation, through one of tenuous relationship to the rest of the "beasts," to one of integration with a vast array of living things in the family of earthly life. Much of our present understanding of human history and relationships is based on analysis of the fossil record, that archive of life's past experiments retained in the earth's crust and revealed to us by the patient efforts of paleontologists, geologists, and evolutionary biologists the world over.

Our history is recorded not only in the fossils of the earth but also in the forms and behavior of living animals. The earliest vertebrates, the first fishes, the amphibian pioneers of terrestrial living, the primordial mammals—all of these live on in our minds and bodies. We mammals—humans, dogs, horses, antelope, anteaters, bats, rats, cats, kangaroos, whales, weasels, and other "beasts"—are all more alike than different from one another in both form and awareness. We owe our success to the countless lives and hard times of our premammalian ancestors, whose stories remain encapsulated in our beings like successive layers in an onion. The ghosts of those ancestors haunt us still, returning in our dreams and manipulating our daily lives across the vast reaches of time in which we were annealed. This book is a look at their history, the grand tale of the class of "beasts" of which the writer and the reader of these words are the handiest representatives.

For some time now, most educated people have agreed that we human beings are, among animals, most logically placed in the class Mammalia with all those creatures who nourish their young with milk produced in skin glands called mammae. The implications of this manner of caring for the young are far-reaching because mammals, unlike the vast majority of the perhaps ten million species of animals on this planet, are born helpless.

Not only must they be fed by their parents, nourished literally from the bodies of their mothers, but infant mammals must also seek maternal protection and, most important, must to some extent *learn* from their parents. Mammals are born with a degree of mental formlessness found in almost no other class of animals save perhaps among the higher birds; in order for the mammal to survive and reproduce its kind, this amorphous infant intellect must be shaped into a useful set of behaviors by the time the young leaves the parental bailiwick. Thus mammals are uniquely flexible in their behavior, generation to generation, and all of them are to some extent able to pass on learned experience to their young—to educate them, to lead them forth into the world (the word *education* derives from Latin roots meaning "to lead forth"). Education is a mammalian forte; the profession of teaching, then, is central to the soul of our kind.

When we think of mammals, we tend to think in terms of bodily insulation —fur usually, or a layer of fat, or clothing. Mammals work hard for a living, and in so doing, maintain a constant body temperature largely independent of that of their surroundings; to this end, various physiological processes and the insulative covering of the body combine to produce a condition known as endothermy (heating from within), or what is commonly called warm-bloodedness.

At present, class Mammalia is a highly diverse group containing some 1,000 or so genera in 118 families inhabiting all landmasses of the earth and most of its oceans. Their enhanced ability to learn, combined with their highly developed system of parental care and their fine-tuned metabolism, has permitted mammals an unparalleled ability to change the face of their planet. This ability is exemplified by the species *Homo sapiens*, Man the Wise. Sitting precariously at the apex of the planetary energy-consumption pyramid as we do, humans too often succumb to the misapprehension that we are the most advanced, "highest" form of earthly life. Then, generously embracing our relatives among the mammals, we claim them all as representative of the way Life Really Ought to Be: furry, warm, and soft.

Whether or not this is so, the story of the mammals' climb to their present status is a long and spectacular one, fraught with interclass warfare and ecologic disaster. Although they really inherited the earth only after the mysterious disappearance of the dinosaurs some 65 million years ago, mammals have been around for much longer, their earliest known remains dating from the late Triassic period, around 200 million years ago. And those early mammals were themselves descended from animals that, were we to see them today, we might think at first glance were not of this earth. They were the therapsids, whose name is coined from the Greek *therion* ("wild animal"), plus the New Latin suffix *ida* ("having the form of").

Often called "mammallike reptiles," therapsids were transitional between more primitive "reptiles," which were the first true land-dwellers among vertebrates, and their more familiar mammalian successors. There are no more living therapsids, with the possible exception of two marginal Australian animals, and we do not miss them a bit. As we will see, most of them would look to us as if the forces of evolutionary selection had experienced a severe case of the jimjams during the therapsid heyday. Nonetheless, therapsids were at one time the most progressive animals that had ever existed; in fact, they ruled the terrestrial world. Their period of glory was the predinosaurian Permo-Triassic interval of between 260 and 190 million years ago. During that vast period, therapsids occupied many of the ecologic niches now inhabited by their mammalian descendants, in a parallel earth that seems in retrospect almost like a satirical reflection of our own. Therapsid "lions," "mice," "deer," "rhinoceroses," and other ludicrous creatures conducted their affairs in a world of plants and animals absurdly different from our own.

As we will see later, some therapsids presaged mammals in many physiological characteristics, such as high activity levels and perhaps furry bodily insulation, to such a degree as to make their transition to true "mammalhood" pretty much a blur in the fossil record. Indeed, although at one time a mammal was pretty much a clear-cut mammal to naturalists, when confronted with the fossil record of our early evolution we are quite unable to discern a

---

Chart illustrating adaptive radiations of land animals discussed in this book. The synapsid-mammal group is bounded by a bold line. Successive evolutionary events are marked as follows: A. Radiation of the Pelycosauria in response to their early improvements in locomotion associated with insect-eating, and, later, the emergence of pelycosaurs capable of eating vegetation. B. Radiation of the Therapsida in response to their improvements on the insect-eating capabilities of small pelycosaurs, and, later, production of a wide variety of herbivores supporting numerous carnivores. C. Fading of the therapsids in response to selective pressure from the archosaurs. Through the rest of the Mesozoic era, the mammalian descendants of the therapsids had to survive by staying out of sight of the swift, sharp-eyed dinosaurs. D. End of the Mesozoic. Adaptive radiation of mammals to fill the ecologic niches for large animals emptied by the extinction of the dinosaurs. E. End of the Cenozoic. Diminution of the mammalian class with the rise of humanity. Major conversion of the planetary biomass into human bodies.

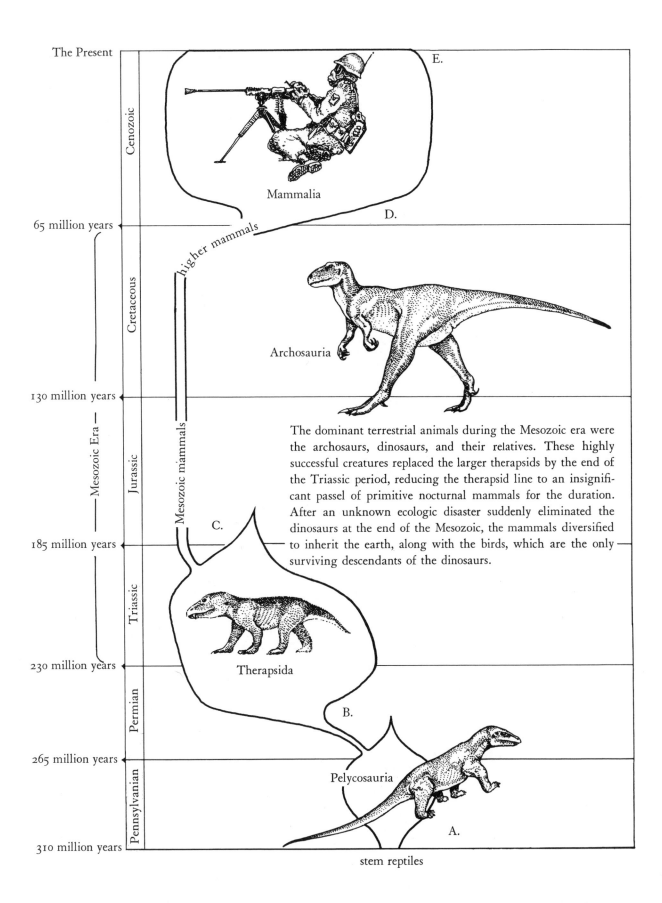

The Present

Cenozoic

65 million years

Cretaceous

130 million years

Mesozoic Era

Jurassic

185 million years

Triassic

230 million years

Permian

265 million years

Pennsylvanian

310 million years

E.

Mammalia

D.

higher mammals

Archosauria

Mesozoic mammals

The dominant terrestrial animals during the Mesozoic era were the archosaurs, dinosaurs, and their relatives. These highly successful creatures replaced the larger therapsids by the end of the Triassic period, reducing the therapsid line to an insignificant passel of primitive nocturnal mammals for the duration. After an unknown ecologic disaster suddenly eliminated the dinosaurs at the end of the Mesozoic, the mammals diversified to inherit the earth, along with the birds, which are the only surviving descendants of the dinosaurs.

C.

Therapsida

B.

Pelycosauria

A.

stem reptiles

sudden leap to the condition of mammalian living as exemplified by the nursing of young—soft structures such as mammary glands are not preserved through the aeons. Because of this inconclusiveness in our history, paleontologists have been forced to set an arbitrary dividing line between therapsids and mammals based on the form of the bones of the jaws; in "reptiles" and therapsids, the lower jaw articulates with the upper through a quadrate bone, while in mammals the lower jaw hinges to the rear of the cheekbone.

Therapsids themselves are descended from an earlier group of "mammal-like reptiles" called pelycosaurs ("bowl-lizards," from the shape of their hip-bones), whose origins date from the earliest attempts by vertebrates to leave the water world for good. Primitive pelycosaurs crawled about the swamps of the Pennsylvanian period around 300 million years ago, during which time they were the most fearsome predators on earth. Pelycosaurs were also among the first land-dwellers to experiment with some sort of control of body temperature, and their progress in this direction led to their founding of, and ultimate replacement by, the more efficient therapsids.

Together, the bizarre pelycosaurs and their therapsid descendants are grouped traditionally in a subclass Synapsida of the class Reptilia. This great subclass owes its early success to one or more of its members' having learned to eat vegetation; in a world of flesh-eaters, these pioneer vegetarians founded the first balanced vertebrate ecosystem—that is, one containing both primary consumers (eaters of plants) and various levels of carnivores feeding on these gentle herbivores. Throughout their time of ascendance, synapsid herbivores—themselves wondrous creatures indeed—fed an astonishing variety of grotesque synapsid carnivores to populate a fantastic world of synapsid diversity that lasted more than 100 million years. In so doing, synapsids set ecologic patterns that would be repeated many times to come; in essence, they wrote the drama of land-dwelling life on the earth.

Because the line dividing pelycosaurs from therapsids is just as fuzzy as that between therapsids and mammals, in this book we will loosely consider their story as the collective tale of one group, Synapsida, a name coined by taxonomists to mean "Those with Fused Arches." We will follow this tale from its inception with the vertebrate invasion of land, through the advent of the first true mammals, examining, as we go, the fall of the therapsids with the rise of the dinosaurs, closing with the setting of the stage for the mammalian conquest of the earth close upon the disappearance of the dinosaurs. In the process we will be tracing the appearance of many of the traits we call "human," including, perhaps, the origins of what some are pleased to call "spirit."

EVOLUTION AND ITS DISCOVERY

It is a commonplace that evolution produces structures only where their function is necessary to the continuation of their possessors' genetic lines: form follows function. Organisms in which for some reason this ceases to be so are selected out from among the living, and die without reproducing. Thus it may be said that all organisms are children of adversity: the family tree is pruned by death. This process of pruning is referred to as selective pressure, the "force" that directs evolutionary change to encourage experimentation among organisms and diversity in life as a whole.

While life itself strives toward stability, or homeostasis, environmental change creates selective pressure requiring change in response. If such change in response does not occur, a line becomes extinct; if suitable change does come about, the new form may experience a spate of adaptive radiation (a rapid diversification in form and function) as it fills-out new econiches in taking advantage of potentials created by the change. Once all the possibilities of a change are exploited by an organism—that is, when new econiches are filling up through adaptive radiation—selective pressure again comes to bear on the organism, this time from its own numbers.

This cycle of pressure–change–adaptive radiation–pressure is central to the evolutionary process, and we will encounter it again and again during our journey through synapsid history. Because the same cycle dictates the history and affairs of human civilization, it is easily illustrated by a cultural analogue, the history of the crossbow.

In human political affairs, there has long existed an "econiche" for hand-held projectile launchers, useful tools in the settlement of disputes not amenable to mere reason. At one time this niche was occupied by the longbow and arrow, which continued in its role virtually unchanged for thousands of years. However, during medieval times pressure arose to increase the efficiency of hand-launched projectiles against hardened targets such as knights in armor. A "mutation" occurred in the form of the invention of the crossbow, far more powerful and more accurate than the longbow.

Everyone immediately wanted a crossbow, and the new weapon quickly became widespread—it "filled up its econiche"—and in so doing, revolutionized warfare. Pressure then arose for diversity: when you have a crossbow and your opponent has a crossbow, it behooves you to increase the range and accuracy of your own weapon in order to nail him before he nails you. Thus "mutations"—inventions—occurred in response to this pressure. Metal crossbows appeared, with far greater power than that of their wooden "ancestors," and sights appeared to improve their accuracy. Simple wooden crossbows were "selected out" as their owners either purchased newer models for their own use or were skewered by others sporting same.

A crossbow, our evolutionary model. This drawing also illustrates the principle of parallelism, in which two organisms (or, in this case, cultural artifacts) subjected to similar selective pressures tend to evolve similar characteristics—note the general similarity in configuration between the crossbow and a modern rifle.

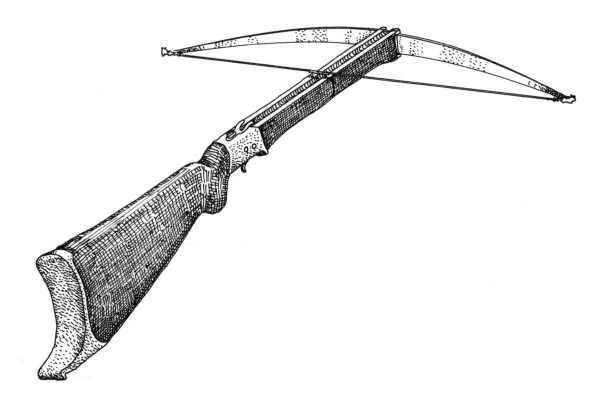

Now everyone had long-range metal crossbows with sights, but anyone who could increase his firing speed had an advantage over the others. Thus pressure arose for "rapid-fire" crossbows. Repeating crossbows appeared, in which five or more bolts were stored in a weapon's stock, to be raised automatically into firing position when the bowstring was set to the trigger mechanism. Everyone then wanted repeating crossbows, and these in their turn experienced an "adaptive radiation."

Sometimes an insurmountable change arises that ends or sharply reduces the variety of a formerly successful line of evolution. This happened to the crossbow with the invention of the chemical gun, or firearm, a "mutation" in hand-held projectile weaponry that was far more efficient at the task to which crossbows had become so abundantly adapted. The crossbow, because of a great adaptive radiation of firearms, became "extinct" as far as organized warfare was concerned. Crossbows function today mainly as curiosities, and as specialized tools for deer poachers, guerrillas, and other hunters concerned more with stealth than with range and power.

In the history of the synapsids we will see analogous cycles again and again. Because its record is so richly documented, the synapsid line is especially well suited for understanding the meshing of ecology and evolution to produce ever more efficient organisms in the face of constant change. Comprehension of this all-important process, while necessary to the future survival of humankind, has not come easily to us; even now, many people reject the idea of evolutionary progression in favor of "human-centered" theories of special creation, geocentric astrology, and the like.

This conservatism is understandable in light of the fact that the study of evolution in all its magnificence is still relatively new to us. Until the last two centuries or so, most philosophers of the West believed that all living things came into being at a single instant (or perhaps during a single week) of creation, formed by a God which was itself a living being of some sort. The naturalists of those days suggested in their innocence that this God had used templates—"bird" templates, "beast" templates, "reptile" templates, and so on—to produce the various groups of organisms with which the world is populated.

One species, *Homo sapiens*, was exempted from such classification on the ground that its supposedly unique faculty of self-awareness was created specifically to mirror the divine mind. Although clearly sharing certain characteristics (constant body temperature, hair, number of limbs, feeding of young with milk, inevitable mortality, and the like) with the "beasts," and like them created of the dust of the earth, the human being was said to have been given, as well, a soul, an unearthly part reflective of the creating God.

Nonetheless, the "bestial" part of human behavior continued to trouble thinking men for millennia. The better to deal with "bestiality," they divided it into categories such as pride, covetousness, lust, wrath, gluttony, envy, and sloth. They then attempted to eliminate this behavior by assigning to it the appellation of "sin," and condemning it roundly—which served, of course, to increase its attractiveness. Thus the human condition was sectioned into sinfulness on the one hand and godliness on the other, and throughout the history of religion, weighty tomes have been compiled on the subject.

In old classifications of the animal world, called bestiaries, animals were said to have been created specifically to illustrate moral precepts useful in the conduct of human affairs. However, with the slow growth of empiricism and careful observation of animal structure during the seventeenth and eighteenth centuries, naturalists began to suspect that similarities between groups of animals were more than a result of godly convenience in creation. From the dissecting table especially, word passed that the construction of human beings and of many beasts was, barring a few specializations here and there, nearly identical. Although the equating of structures in humans and beasts was considered lewd and sacrilegious by much of the polite society of the times, evidence mounted that some sort of progression of changes could be traced in the forms of extinct and living animals—a progression that might include humans in a continuum of structure and function common to animal life as a whole. Threatened with a loss of its unique place in the special creation, much of educated humanity recoiled in fear and disgust.

Nevertheless, the data being amassed by the infant natural sciences continued to force the walls of philosophic reaction. The publication in 1737 of Carl von Linné's *Systema Naturae* marked the first classification of human beings in a class Mammalia of "beasts," including all animals whose young are suckled with milk. Von Linné further insulted his contemporaries with the salacious suggestion that humans might be associated closely with monkeys and apes, by establishing the order Primates (from the Latin word for "first") to contain these creatures at the apex of class Mammalia.

The latter half of the nineteenth century witnessed evidence presented by Charles Darwin, Thomas Henry Huxley, and others that humans were actually *descended* from other animals, in a process beginning at some single point in the remote past and stretching through the aeons and through many branchings and convolutions to produce the dazzling variety of living things with which our planet is populated today. No longer could thinking human beings rationally regard themselves as separate from the rest of the natural world and its laws (although certain *un*thinking humans, notably politicians and economists, continue to do so).

In the light of a growing collection of fossil animals from around the world, our history is plainly one with that of the other "beasts." It is a long story; the synapsid-mammal line alone is some 300 million years old, that of the vertebrates perhaps 470 million, that of life itself more than 3.5 billion. So, although "bestial" behavior certainly troubles us at times—and could even kill us off—we may be assured that it comes naturally to us!

ROOTS

We mammals and our relatives among the synapsids are, above all, vertebrates, animals whose central nervous systems receive independent protection from a series of rings of bone or cartilage, called vertebrae. Vertebrates made their entrance during the Ordovician period around 470 million years ago, and it is in that remote time that we will begin our journey through the history of our kind.

A human visitor to the Ordovician would find our planet a very different place—unearthly, we might say. Although the atmosphere might be breathable, having already gained a large oxygen content from the photosynthetic activities of primitive marine plants, the Ordovician continents were devoid of life. Lonely winds whispered across the bleak rock and sand of land as deserted as the chill waste of Mars; no touch of living green graced the harsh, storm-blasted terrain, no bird or beast or insect moved about. Indeed, the Ordovician landscape in some ways resembled the urban northern end of the state of New Jersey today.

At the water's edge, however, Ordovician life prospered and danced as enthusiastically as does tide-pool life today. For more than 3 billion years, life had been experimenting with new ways of supporting itself in the earth's oceans. Fueled by the photosynthetic work of a host of water-dwelling algae, life had already differentiated into many of the groups still characteristic of aquatic life today. Some of these were sessile (stationary) animals living by filtering floating edibles from the water and protected from predators by shells or burrows. Others, especially of the phyla Mollusca and Arthropoda, had evolved quick movements and acute senses in order to prey on their slower cousins. It was the presence in Ordovician seas of these active animals that provided the impetus for the evolution of the chordate plan and the birth of the Vertebrata. We backboned creatures are children of the Ordovician, conceived more than 400 million years ago among progenitors hard pressed by invertebrate predators related to modern clams and insects.

Chordates are marked by three adaptations originally useful in the escaping of active predators: a stiffening rod, or notochord, for which they are named (from the Greek for "back-rope"); pharyngeal clefts (slits in the throat), through which water may be conducted for breathing; and a dorsal nerve tube extending along the back, through which activity may be coordinated by

---

Hemichordates. Although the adult is a wormlike animal living in a hole in the mud, the larval form is a free-swimming ciliated animal. Retention of free swimming in adults led to the first vertebrate adaptive radiation.

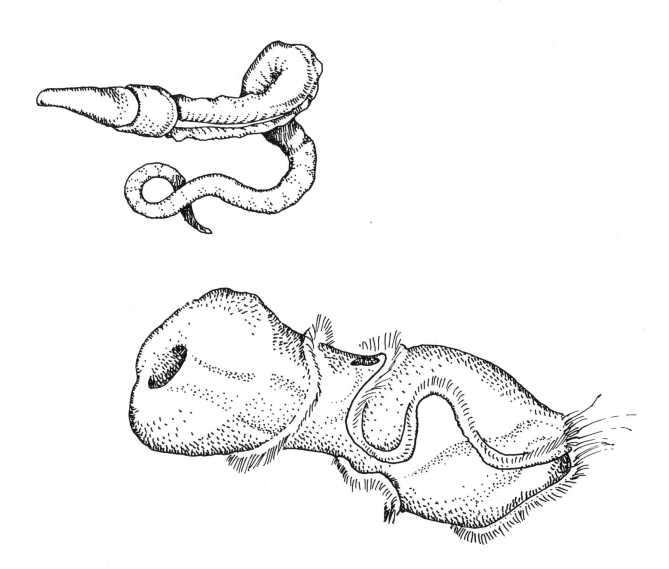

means of a central nervous system. Each of these adaptations reflects pressure to flee from predators, and all of them may be seen in certain very primitive living relatives of ours, the most familiar of which are small marine animals along the lines of lancelets of the genus *Amphioxus*.

Because our prechordate ancestors were boneless things, their fossils have not survived the intervening ages to show us our precise origins. However, specialized offshoots of the primitive chordate line survive today in the form of members of a phylum Hemichordata ("Half-chordates"), including two minor classes of wormlike animals. Modern hemichordates live either anchored to the sea bottom or in burrows in the marine mud. They are uninspiring animals, some of which filter microscopic food particles from the water, others of which glean their eats, earthwormlike, from the organic matter in mud. Like the earthworm, but independent of it, hemichordate ancestors had evolved an elongated, bilaterally symmetrical (two-sided) form. They certainly do not resemble us closely, nor would many of us be pleased to acknowledge one at a family reunion. Nonetheless, we should revere the memory of these humble ancestors, from which would ultimately descend such unlikely creatures as sharks, therapsids, and the President of the United States.

It is among the larval, immature hemichordates that we find likely candidates for our ancestry. These small, free-living creatures swim about in search of permanent homes by means of ciliary bands equipped with thousands of waving hairs. In living species, such larvae serve as dispersal agents for their kind, swimming for a time before taking the sessile adult form. However, during the Ordovician period, arthropods and mollusks learned to dig the immobile adults out of the mud; the resultant selective pressure produced in some hemichordatelike animal a mutation that resulted in a new form whose adult stage retained the ability to swim. This type of mutation, which adults retain infantile characteristics, is called neoteny and is a relatively common event in the process of evolution.

As primitive chordates improved on their swimming ability to escape the disconcerting array of shelled squidlike mollusks and giant scorpionlike arthropods with which they had to share their neighborhoods, they gradually abandoned their ciliary swimming organs in favor of a new tail of muscles playing along the notochord to produce leverage against the surrounding water. Differentiation between the functions of the forward (mouth) and after (tail) ends of such animals gradually resulted in a process of encephalization, or the forming of a true head containing most of the major sense organs, another product of the development of swimming speed. After all, if

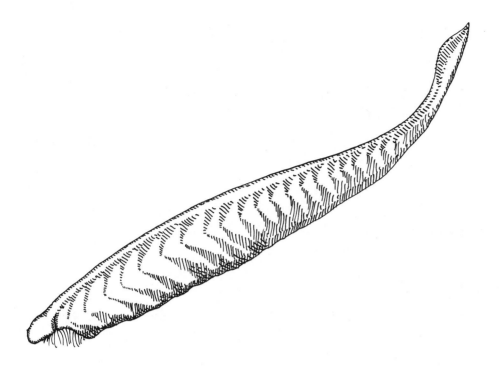

Amphioxus, a primitive chordate whose structure presages the fishlike form that was to lead to all higher vertebrates.

---

an animal is to move quickly, it had best know where it is going before it gets there, and what it is going to meet when it does.

To this end, the forward part of the nerve cord evolved swellings to process sensory input; at the leading end of the cord was a center for chemical analysis of water conditions, better known as a sense of smell. Immediately aft of this sniffer center was one that processed information relating to light conditions in the water; this gradually improved to become what we call an organ of vision. Behind the vision center was one constructed to process information relayed by vibrations in the water and to orient the animal right side up. These three parts of the primordial chordate brain are called, respectively, the prosencephalon (forebrain), mesencephalon (midbrain), and rhombencephalon (hindbrain); and their order provided the framework on which was to be built the brains of all higher chordates. Although it cannot be denied that these early chordates were lacking in intellectual profundity, we must remember that there was little in their lives to require same; their routine consisted of seeking slow-moving food, escaping predators, and occasionally mating in a desultory way. Having no limbs, these first chordates did not have to make even such simple decisions as which might be the best foot to put forward.

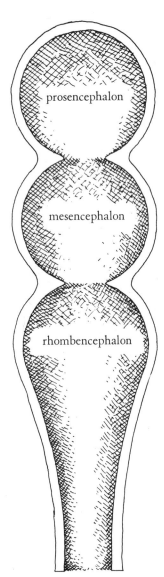

The original triple-bulged vertebrate brain, a simple enlargement of the fore-end of the nerve cord.

prosencephalon

mesencephalon

rhombencephalon

Because of the importance of the nerve cord in integrating the activity of early chordates, selective pressure again arose, this time for some sort of protection for this communications line. By the middle of the Ordovician this had been accomplished in the evolution of a vertebral column, in its simplest form a series of rings encircling and armoring the nerve cord and ultimately obviating the necessity for the older stiffening notochord. With the invention of the vertebral column the chordates could slow down a bit, evolving armor with which to protect themselves. This was useful because these were still jawless beings with no appendages for catching or chewing up meals, and were thus still dependent on the goodies they might find in the mud at the bottom of the water.

These agnaths ("jawless ones") were so completely specialized for bottom-dwelling that they had evolved three eyes: two for spotting food and predators to either side, and a third, central or pineal, eye for detecting the swoop of predators from above. This third eye persists in descendants of the early vertebrates, a reminder of our mud-grubbing days that still functions as a light receptor in some modern reptiles. While in more progressive animals such as mammals and birds it is completely covered over, the pineal body still functions as a timing device for hormonal activity based on seasonal or hourly light changes. During the time of the synapsids this organ played a significant part in their evolution.

The ancestral vertebrates were finally to rise up and overthrow the dominion of arthropods, to become formidable predators in their own right. First, however, they had to overcome their main impediment in feeding, the jawless structure of their strainerlike, mud-filtering mouths. We have seen that chordates possess gill slits, the neck openings through which water exits after

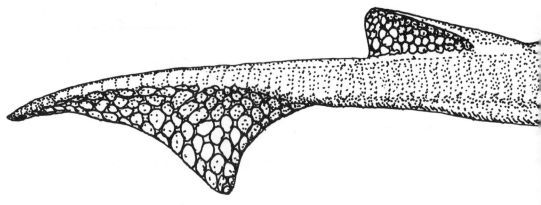

entering the mouth and flowing across the gills. With the advent of vertebrate status and the evolution of a skeleton, the gills themselves became supported by bony arches occurring in a series along the gill slits. Early in the history of the only extinct vertebrate class, the Placodermi ("Plate-skinned" Fishes), the foremost of these arches took on the function of jaws, the upper half of each side of the gill arch articulating with the lower. With this innovation the vertebrates exploded across the watery world of the Devonian period around 410 million years ago in a spectacular adaptive radiation that is aptly called the Age of Fishes.

From the Devonian onward, vertebrate jaw structure dictated the life-style of its possessors. From barracuda to Howard Cosell, vertebrates depend on their mouthparts to manipulate their environments, to capture energy, to communicate. In the fossil record, it is often the jaws and teeth that show us how these long-gone characters conducted their affairs. The scientific names of fossil vertebrates often reflect the importance of their jaws and teeth by including a syllable such as the suffix *-don*, from the Greek *odous* ("tooth"). We will stumble over this one a lot during our consideration of synapsid history.

Selective pressure from the adaptive radiation of jawed fishes forced some of them to flee to the edges of freshwater systems, to shallow streams and ponds where they were occasionally subject to seasonal fluctuations in the

---

A primitive vertebrate, whose jawless mouth restricted it to filtering the mud at the bottom of the water. Such animals may have been equipped with electric organs similar to those in modern electric fishes, the only means of defense in such slow-moving animals besides their armor.

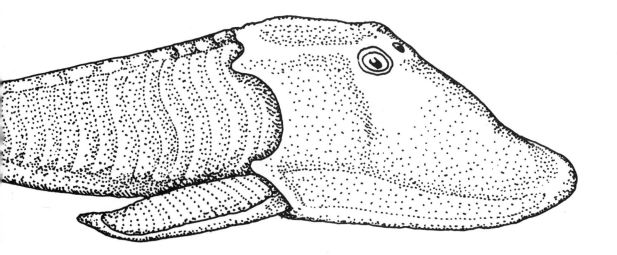

water level. Here their descendants remained marginal animals adapted to pushing themselves around with sturdy, fleshily muscled fins through expanses of shallow water, occasionally gulping a breath of air into a lung or two. Such lungs, originally little more than air bladders within swamp-dwelling fishes, originated in response to selective pressure exerted by an occasional lack of enough water to hold dissolved oxygen. It was among these unassuming forms that selective pressure again pointed its fickle finger with the rise of the Tetrapoda ("Four-footed Ones"—including our two-footed selves), the great vertebrate experiment in terrestrial living.

TRANSITION

For aquatic animals, the colonization of land presents problems rather akin to those that would face us should we ever decide to move to another planet. Even gravitational force is different on land, for water-dwellers are used to an essentially weightless environment where the density of the surrounding water approximates that of the animals themselves. Therefore, aquatic animals need expend almost no energy in supporting their masses, the water doing all that for them.

Animals making the transition to terrestrial living, however, must be prepared to give up this easy life, for landlubbing requires some sort of locomotor apparatus to take care of bodily support and movement, plus more subtle changes in internal structure to keep organs from being collapsed by gravitational structure. We can get a look at what happens to a water-being on transition to dry land by examining a newborn baby. Such a mite cannot even hold its own head up, and must be enfolded in adult arms so completely as to simulate the watery world from which it has just emerged.

Also, of course, the terrestrial atmosphere is much thinner, lighter, and drier than the medium in which aquatic animals are used to getting about. This presents problems in moisture conservation that must be surmounted by changes in the body covering if pioneer land animals are not to dry up and blow away. Moreover, the surfaces of vertebrate eyes must be kept lubricated to a certain extent to permit good vision; when the animal moves to land, some mechanism must be arranged to protect these delicate eyes from drying.

Again, temperature changes on land are both faster and broader in range than those occurring in water: the daytime highs are much higher, and the nighttime lows much lower. For animals such as primitive vertebrate swamp-dwellers, used to relatively even and cool water temperatures, fluctuating land temperatures presented a grim obstacle for which they were poorly prepared.

Finally, the reproductive process requires the presence of moisture in which exchange of genetic material may take place. Water-dwelling vertebrates ordinarily fertilize their eggs in open water, and the young in these eggs are moistened by the surrounding water until they hatch. The threat of drying to immobile eggs and immature young was perhaps the greatest obstacle the vertebrates had to overcome in colonizing the land.

With all these problems in mind, it is hardly surprising to us that of all the phyla of animals, only a few ever successfully made the transition from aquatic life to a fully terrestrial existence. Chief among these are the arthropods, such as insects, and our own phylum, the chordates. These groups share a constellation of adaptations that appeared early in their aquatic ancestries and happened also to predispose them to efficient land-dwelling.

The first of these characteristics was the rigid framework or skeleton possessed by both arthropods and vertebrates, offering support for body tissues out of the buoyant watery medium. Secondly, arthropods and vertebrates are relatively active animals, specializing in a certain efficiency and muscularity of movement that readily lent itself to modification for contending with the force of gravity. Finally, both arthropods and vertebrates are equipped with a set of acute sense organs appropriate to animals specializing in dash and grab, organs of balance and sight with which they might anticipate the new problems of gravity presented by the terrestrial world.

Sometime during the Silurian period, between 425 and 413 million years ago, plant life had evolved a capability for surviving temporary recessions of the water level, perhaps along tidal shores or near swamps where occasional drying exerted selective pressure on them to do so. Once these early plants made a complete transition to land-dwelling through the evolution of materials (cellulose and lignin) for support and prevention of drying, they took advantage of a world devoid of grazing animals to diversify in a mighty vegetable adaptive radiation. Within a few million years, much of the earth's land surface was covered by great forests of plants, whose evolution was temporarily uninhibited by the activity of hungry herbivores.

The animal world was quick to follow suit, however. Arthropods appear to have been the first animals to invade the terrestrial environment, probably in the form of amphibious creatures such as isopods, whose descendants survive today in the form of the familiar sow bugs and pill bugs of house and garden. While these crustaceans never fully adapted to a life of land-dwelling, their relatives the scorpions became one of the first animal groups to succeed in making the transition from water to land and to become fully terrestrial. These interesting animals were so efficient and well defended that they survive nearly unchanged today after more than 400 million years.

The animals on which early scorpions preyed were arthropods specialized to feed on the plants at the water's edge. Their descendants, to escape the pressure exerted on them by scorpionid predators and better to exploit the multitude of econiches opened up by the evolution of land plants, took to ever more rapid locomotion and a swift spread inland. Although the more persistently primitive forms possessed many pairs of legs (and hence are often called myriopods, "many-leggers"), some of their descendants during Devonian times reduced the number of their legs to six and evolved wings, subsequently becoming the dominant and most successful animals on the planet, the mighty class Insecta. Thus the terrestrial ecosystem was enriched and complicated as arthropods and plants coevolved to fatten the life of the continents.

Meanwhile, back in the water, vertebrate evolution continued apace. Selective pressure exerted by swamp-dwelling forced the fleshy-finned, air-gulping inhabitants to retain and elaborate on their primitive lungs to cope with the unpredictable nature of their water supply by crawling painfully across land when the water levels fell. In so doing, some Devonian forms inevitably dis-

---

A sarcopterygian ("flesh-finned") fish, *Eusthenopteron*; sarcopterygians lay close to the ancestry of all land-dwelling vertebrates. This one is shown pushing itself slowly across the mudflats of its Upper Devonian habitat with its muscular fins. Although this was certainly a difficult way to make a living, it offered plenty of opportunity for advancement.

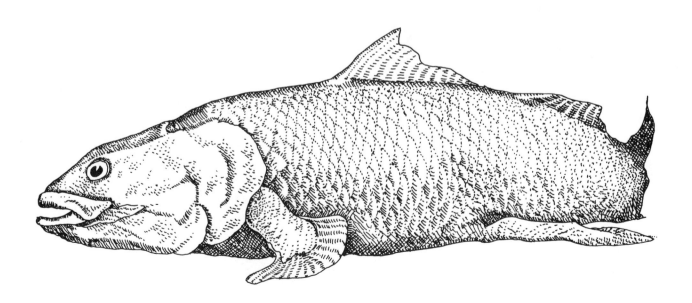

covered that the land was already populated by a rich flora and fauna, an ecosystem concentrating a vast supply of space and energy out of reach of previous vertebrate exploitation. In contrast, the water world was full of vertebrates that had been working for ages to perfect their various little acts, gradually filling econiches and gathering in the energy provided by water plants.

Many of the swamp-crawlers, all of which appear to have been carnivorous, were forced to eat arthropods at the water's edge rather than return to the deeps, for their relatives were cleaning up all the goodies there. Selective pressure mounted for such marginal creatures to spend more time on land, and with suitable modification of their muscular fins, they did just that. Bones in the fleshy part of the fin were reduced in number to become three, jointed flexibly to one another, while the rays became five in number, evolving flexibility and a set of pointed tips—claws—with which these animals might hook the ground for better purchase. This arrangement has persisted through 400 million years in the pentadactyl (five-fingered) limbs of the tetrapods.

The earliest land-dwelling vertebrates had no flash or pizzazz. They were a

---

Diagrammatic views of the origin of tetrapod limbs in the fins of sarcopterygian fishes. The hatching patterns illustrate homologous bones—those whose identity remains the same as evolution proceeds.

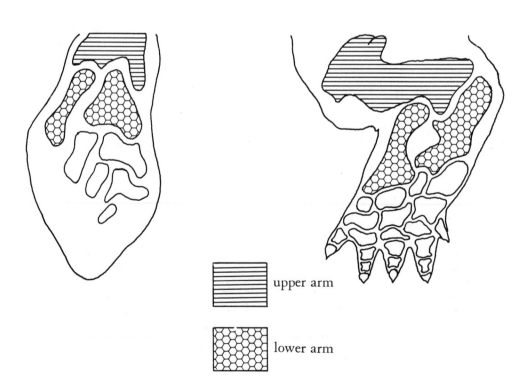

upper arm

lower arm

slowgoing crew, little removed from their fishy ancestors in appearance or, in spite of their new legs, in locomotion; their long, squat bodies still moved in an undulating, fishlike "swimming" motion as they lugged themselves around. Indeed, although doing much of their moving about on land, these early tetrapods were still not at all land-dwellers in one respect: their mating was still confined to water because of the inability of their reproductive cells to meet and survive except in water.

The young of early tetrapods were hatched in water at a very early period of development called the larval stage, and they lived and looked much like fish until at a later stage they developed the limbs and breathing apparatus necessary to make the transition to land. This arrangement persists in modern frogs, salamanders, and other amphibians, and is responsible for the class name of these primitive tetrapods, Amphibia ("Double-lived").

Because their reproductive cycle required them to remain near water, early amphibians were sharply limited by way of interesting things to do. Their bodies were still susceptible to drying, and they appear to have spent much time loafing in the water of vast, warm, deliciously oozy swamps. The structure of their teeth suggests that they were mainly predatory animals that sat around until some smaller creature passed in front of their long, flat heads, at which point they would flop forward and snap in a desultory way.

Potential econiches for crawly, amphibious tetrapods were quickly filled as these spread through the swamps of the Devonian and Carboniferous periods between 410 and 265 million years ago. Pressure increased, especially on small insect-eaters that were too little to compete successfully with their bigger, stronger cousins, and some such animals were gradually forced inland. Their small size and comparatively light builds permitted these little animals a better chance at the insects on which they subsisted; the fact that insects were the fastest animals of their time exerted considerable pressure on the tetrapods to move more quickly, and as they did so, their opportunities inland increased. In the swamps, selective pressure was continuously exerted on the insect-eaters toward some mode of reproduction independent of water. Each time they returned to mate, something might eat them—a poor climax to any lovemaking event and an impediment to the survival of the line.

It is not surprising, therefore, that the small insect-eaters evolved a system of reproduction independent of open water. This came in the form of the amniote egg, in which moisture for the young is enclosed in an amniotic sac, and sac, embryo, and a food supply are enclosed within a hard outer membrane, or shell. Such eggs are fertilized while still inside the mother, protecting the male reproductive cell (sperm) from contact with the air. The whole egg was encapsulated while still in the mother's body, then deposited and probably buried in a moist place to hatch.

The "pond" enclosed within the amniotic sac is called amniotic fluid, and the membrane forming the sac is the amnion; hence the style of egg is amniote. The immensely successful descendants of the tetrapods who pioneered its engineering—the reptiles, archosaurs, and the synapsid-mammal line—all are called amniotes, and all owe their gorgeous diversity to the hard-pressed insect-eating parents of that leathery, saclike egg of 300 million years ago. With its appearance the amphibians sealed their own fate, for the success of their amniote descendants was to force them to the margins of terrestrial ecosystems, little frogs and toads and salamanders in the world of eagles, lions, and human beings.

Note well the fact that insects fueled the early amniotes. The eating of nasty

little six-leggers has resulted in many an evolutionary revolution among our ancestors, for insects change faster than any other land animals. The cause of their unique success is this high-speed genetic mutability, and animals specializing in eating them must keep up. At the same time, however, insects are tremendously abundant, as any camper will attest, and offer lots of food to support a great diversity of bug-eaters from which mutations may be selected as well. Insects are in many ways the most progressive living things; where they lead, others follow. Remember this the next time you mash one.

A seymouriamorph (from Seymour, Texas, where its fossils are found) amphibian shows an ability to lift itself off the ground a bit, which enabled it to get about on land somewhat better than could its amphibian contemporaries. From this or a similar Lower Pennsylvanian form descend all the real land-dwelling vertebrates, the amniotes.

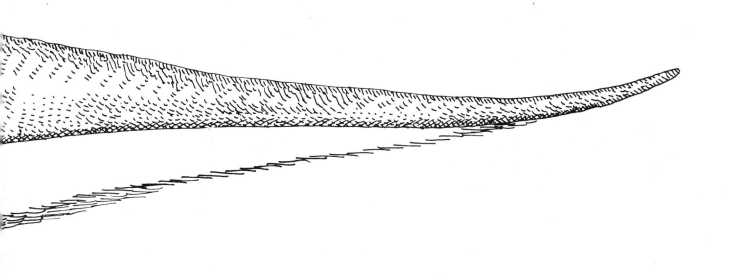

COTYLOSAURIA

Taxonomy is the science (or, in some cases, the art) of classifying living things into groups called taxa based on their evolutionary relationships to one another. As all sciences are, taxonomy is limited by the extent of prior knowledge on which it is based. It is also a highly conservative science in which the judgment of people long dead often restrain the actions of its living practitioners. As we have seen, modern taxonomy owes its birth as such to Carl von Linné, and von Linné's prejudices continue to trouble the functioning of modern taxonomy, exerting their ghostly pressure across two centuries of increasing knowledge.

Thus we are haunted today by the Reptilia, a class of amniotes created by von Linné to include the living crocodilians, turtles, lizards, and snakes with which he was familiar. Reptiles are commonly regarded as those amniotes whose body temperature is determined by that of their surroundings (ectothermy), rather than by internal physiological processes (endothermy) as in mammals and birds. Living reptiles, because of their relatively unsophisticated metabolic systems, are unable to stand erect for any great length of time and are therefore usually thought of as a class of creepy characters; the word *reptile* itself is derived from the Latin *reptilis* ("crawling"). Their ectothermic metabolism has earned reptiles the misnomer "cold-blooded," and because their body workings do not support so highly variegated a set of behavior patterns as may be found in mammals and birds, reptiles are often regarded as stupid, automatic creatures with little recommending them to polite society. In fact, *Webster's New Collegiate Dictionary* offers as one definition for *reptile*, "a groveling or despised person."

Systema Naturae's definition of reptiles was satisfactory, perhaps, for the living forms, but during the ensuing centuries and the birth and development of the science of paleontology, many fossil bones have been discovered with which von Linné could not have been acquainted. Considered in their simplest and most visible aspects, many of these bones appeared to have been designed on a plan similar to that of living reptiles. If this sort of reasoning were used for living animals, even birds and mammals might be (and sometimes are) called "glorified reptiles." However, early paleontologists had little or no conception of the process of evolution and were forced to believe that such extinct forms *were* reptiles. Class Reptilia, then, was conveniently enlarged to include these newly discovered forms, notwithstanding their immense diversity, and a bone-based definition of reptiles emerged: those tetrapods with a single occipital condyle (the joint by which the neck articulates with the skull—as in birds, for example), a well-defined quadrate bone providing a joint between the skull and the lower jaw (also as in birds), and certain details of the ribs and breastbone combining to delineate the class as represented by living reptiles and as many extinct forms as early paleontologists could possibly squeeze in.

Because of their inclusion in class Reptilia, many of the extinct animals to be considered in this book have been saddled with the reptilian image as defined by the living or Linnaean reptiles: the turtles, snakes, lizards, and crocodilians. Thus these long-gone creatures are still usually considered a slow, crawly, "cold-blooded" lot of evolutionary failures. However, with the coming of age of both paleontology and taxonomy, and the application to these sciences of the methods of other life sciences, notably physiology and ecology, we are coming to suspect that a large portion of these extinct "reptiles," if viewed in the flesh, would probably never have been included in that oft-disparaged class. Indeed, class Reptilia shows signs of not existing at all in reality, having served instead as a sort of taxonomic wastebasket for varied extinct amniotes that did not fit in with the Linnaean scheme of things.

In its place we see a tremendous adaptive radiation of amniote forms that began during the early Pennsylvanian period, about 310 million years ago, and that continues to this day. During that vast interval of time, entire ecologic worlds have been built and destroyed, based on various amniote successes and failures. We ourselves are a brand-new amniote experiment, and although our chances for real success on the evolutionary time scale are questionable at best, we can certainly profit from an examination of the trials and errors of our cousins long past.

The stem of the amniote family tree is composed of an order called the Cotylosauria (literally, "Stem-lizards"), a group of low-slung, elongated ani-

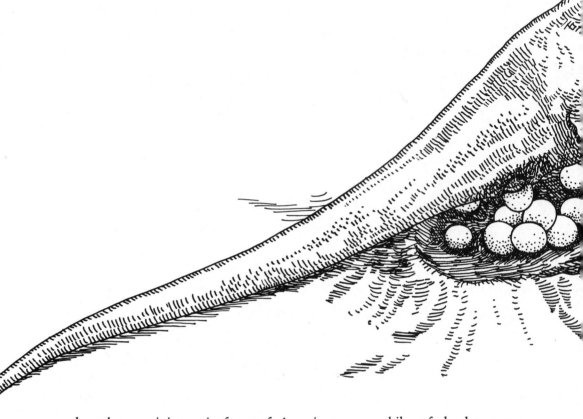

mals rather reminiscent in form of American automobiles of the late 1950s and little removed skeletally from their amphibian ancestors. Although some of these reached nearly two meters (about six feet) in length, the majority maintained a smaller size in response to the exigencies of fighting gravity and chasing insects. Almost all cotylosaurs were predators, eating arthropods, one another, and whatever amphibians they could catch. Some of them even returned to the water when hatched, having avoided the vulnerable aquatic larval stage, to prey on the relatively defenseless young of amphibian swamp-dwellers and hasten the demise of the old order.

The tremendous advantages conferred on the new amniotes by their uncontested invasion of the terrestrial uplands resulted in immediate adaptive radiation as new econiches presented themselves. With each conquest of a new niche, special adaptations in form and function were required and diversity increased. As is always the case in a successful radiation, the descendants of the triumphant innovators came into conflict with one another as they filled newly available niches, and they then began to exert pressure on each other all over again.

Hylonomus ("Wood-dweller"—its fossils were found inside those of Lower Pennsylvanian hollow logs), one of the earliest amniotes. Actually, the division between animals with water-dependent amphibian eggs and those with dry-land amniote eggs must remain moot, for we have no record of the eggs in conjunction with the animals that laid them. *Hylonomus* was an insect-eater whose mode of life called for greater activity on land than was possible for amphibians. The insect-eating impetus for vertebrate evolutionary breakthroughs is a dominant theme in our history.

One early result of this pressure among the cotylosaurs was the development by some of them of enlarged ribs that served to protect them from being crushed by the jaws of their bigger relatives. This defensive ploy was so successful that their descendants perfected the new body armor to survive some 260 million years to the present in the form of turtles and their kin, the Chelonia. Because of the mass and rigidity of their enclosing armor, turtles have always been limited in the speed with which they can get about on land, a fact often exploited by fabulists. In water, however, they can move swiftly enough to be effective predators, often achieving large size.

Their armor having so successfully released them from much of the trouble caused by predation, turtles have remained an extremely conservative group, much as the rich among human beings, armored by their wealth, often do. Once this armor was perfected, turtles seem to have seen little cause to change their form; "slow but sure wins the race," but at the cost of any kind of mental or physical oomph beyond that of their postcotylosaurian ancestors. As a matter of fact, modern turtles give us a good look at the state of vertebrate evolution shortly after our ancestors learned to reproduce out of the water; in at least gait and personality, the turtle accurately recalls our common progenitor somewhere back in the dim reaches of the Pennsylvanian period.

As noted before, temperature regulation is a major concern of terrestrial animals, as any New Englander's winter oil bill will attest. Each land animal must maintain its body temperature within a certain narrow range in order to function efficiently. Because watery environments change temperature relatively slowly, their habitants are able by means of patient, slow physiological accommodation to cope with most such variations within the narrow range of their adaptations. When temperature drops too low, ectothermic water-dwellers such as amphibians simply shut off, falling into a torpor until the next increase in temperature. As temperature rises, such animals become more active; however, because most bodies of water on the earth seldom get much warmer than about 30° Centigrade (around 85° Fahrenheit), amphibians tend to function most efficiently at this temperature.

When our small insect-eating cotylosaurian ancestors began wandering farther from water, they experienced rigid selective pressure from the capricious temperature changes of the air around them. Between day and night, air temperature might vary 20°C between extremes well above and below those experienced by amphibians living in water. While early "Stem-lizards" probably persisted in becoming dormant in cool weather, the higher daytime temperatures on land required special accommodation if the pioneer land-lubbers were to survive.

Some of them probably coped with this problem by increasing their temperature of most efficient functioning to take advantage of daytime temperatures of up to 40°C (about 104°F), which were common in their tropical habitats. From this advance they derived an added benefit, in that higher body temperature permits more rapid metabolism and thus a more efficient use of food energy for muscular activity—in this case, chasing insects. Such a condition is found today in the small insect-eating lizards of the desert American Southwest, tremendously fast and agile little animals whose body temperature, while still dependent on their surroundings, is comparable to our own (about 38°C).

Early high-temperature animals were confronted with new difficulties of temperature control. Fat animals, for instance, have less body surface (compared to body volume) through which heat may be transmitted than do small animals; therefore their body temperature changes more slowly than does that of small animals. On cool mornings when their cold bodies heated slowly, these bigger specimens might be nibbled away by smaller, faster-heating relatives, a very inconvenient situation exerting considerable selective pressure toward small size at the beginning.

Small animals, while able to exchange heat quickly with the environment through their skins, were subject to sudden cooling when the temperature dropped, as in the evening. This could immobilize them, and fatter relatives, whose temperatures changed more slowly, might snap them up like Animal Crackers. Needless to say, pressure was strong toward temperature control of some sort. It is hardly surprising, then, that pioneer land-dwellers quickly began to deal with this problem in a variety of ways—ways that were to have a profound effect on all the future history of life on the earth.

Many early amniotes took the route of evolving behaviors to cope with daily temperature changes. Ancestors of the living turtles and other reptiles became so adept at seeking the sun for warmth and shade for cooling, and at becoming dormant during times of intense heat and cold, that their metabolisms have changed very little in the intervening millions of years. While turtles appear to have done so partly by keeping near water, many other primitive landlubbers evolved fine-tuned control mechanisms based on detection of subtle environmental clues to warn of impending temperature changes.

Remember that third eye in the top of the head of primitive vertebrates? Well, this organ survived among the first amniotes, and in many of them it became quite large. However, on land its old function, that of detecting the swoop of swimming predators from above, could not have had any importance because the largest flying beings in those days were insects of various sorts. Therefore, we are led to believe that the pineal eye in land animals was mainly a sensor of conditions in the sky—such as the rhythms of the sun—that might bear on the well-being of the strange three-eyed creatures attempting to populate the continents.

Experiments with modern lizards possessing the pineal eye tend to bear out this conclusion: if light is blocked from the central eye, the animal's ability to regulate its temperature behaviorally is impaired. In the remote days of the Pennsylvanian and Permian periods those terrestrial animals of slower movement had particularly large pineal eyes; this suggests that the third eye in these animals was an important warning device to ensure anticipation of environmental change so that these unwieldy creatures might prepare in advance.

Conversely, in more advanced animals such as mammals and birds, whose body temperatures are controlled internally, the pineal eye is gone; a remnant survives in the pineal body within the brain, which still serves to coordinate seasonal and daily behavioral changes in response to environmental conditions. Even human beings possess a pineal body, whose exact effects on consciousness remain unknown—we can't simply take someone's pineal body out, at least not without the person's permission, to see what happens. However, this ancient organ may still have great sway over our behavior; perhaps some inkling of its importance was guessed at by the ancients, who dubbed the pineal body "the seat of the soul," although modern research suggests, as we will see, that the soul is more closely related to the nose.

All this insect-eating by little lizardlike cotylosaurs could not fail to have some effect on the insects themselves, which responded to this pressure by producing a variety of much faster, more agile creatures. During the Pennsylvanian period of about 300 million years ago, insects experienced a great adaptive radiation based on improvements in locomotion, and, naturally, selective pressure was exerted on those little insect-eating "Stem-lizards" to improve their own means of transport.

Increase in speed requires greater energy to fuel it; increased energy, if you are an insect-eater, requires more speed for capturing insects. The increase in speed toward which "Stem-lizards" were being pressed necessitated, among other things, a more efficient gait with which they might run after insects, plus a more effective snap of the jaws with which they might seize their prey. Improvement in these areas led to larger catches of insects, which exerted pressure on the insects to increase their own speed, which turned more pressure on the little vertebrates to do likewise. This push toward more active living produced some important innovations in both locomotion and the method of biting, innovations that were to create an entirely new way of insect-catching—and, ultimately, an entirely new kind of animal. So, though many readers no doubt dislike insects and their creepy little arthropod allies, we owe our all to their intensely creative abundance. In the pursuit of insects were born the Synapsida, our remote ancestors, and to synapsids we now turn for one of the most phenomenal evolutionary tales our planet has produced in over five billion years.

ENTER THE SYNAPSIDA

During the Pennsylvanian period we find several different sorts of amniotes whose skulls already show a tendency toward use of the higher operating temperatures and greater appetites to which their kind was becoming accustomed. This tendency is reflected in an increasing jaw length, and thus an increasing gape, and in muscular adaptations to permit this big-mouthed bug-munching crew better purchase on their armored insect prey. The various adaptations facilitating insect-eating permit us another digression into the weird world of taxonomy and a look at the roots of all the higher amniote groups inhabiting the world today.

The skull roof of early amniotes, little removed from their amphibian ancestry, was completely covered over with plates of bone. The muscles operating the jaws of most of these uninspiring animals were concealed beneath the bones of the rear of the skull, attached to the inner surfaces of these bones and serving to operate comparatively narrow gapes suitable for engulfing very slow animals such as larval amphibians or worms. With the chasing of insects, new arrangements became necessary to increase the speed of grab and strength of pinch. (Among the prey of the primordial landlubbers were just such cockroaches as live behind sinks in nice people's houses today. And while most of these nice people have progressed somewhat beyond cockroach-eating, they still have occasion to chase them and may thus experience one of the selective pressures that operated on our remote ancestors in the swamps of the Pennsylvanian.)

Since running was as yet imperfect among vertebrates, most of the initial insect-eating response occurred in the structure of the skull, in which larger spaces gradually became necessary to accommodate larger, more efficient jaw

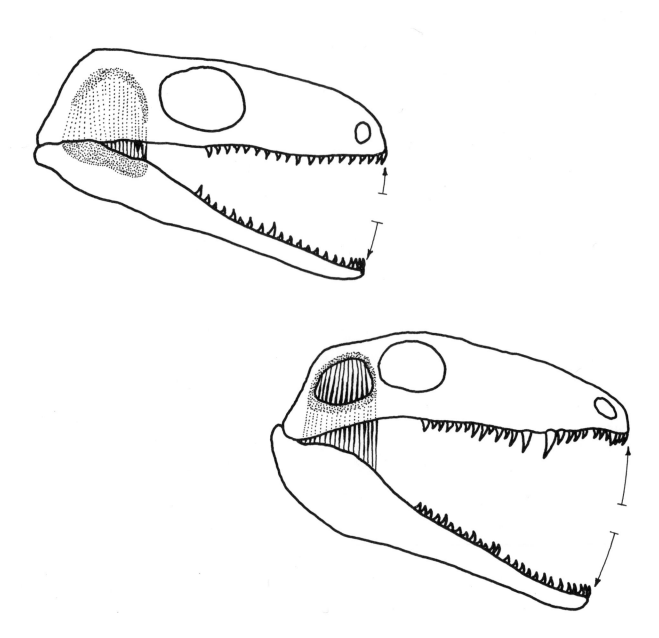

Anapsid ("archless") and apsid ("arched") skulls. The temporal opening in the arched model (*below*) permits greater muscular stretching and bulging, and hence allows a wider gape.

muscles; this process eventually resulted in the appearance of fenestrae (windows), or holes, in the sides of the skull back behind the eye sockets. The edges of these fenestrae provided additional attachment surfaces for the increasingly strong jaw muscles, while the openings themselves permitted these muscles to bulge when closing the jaws.

Several groups of early land animals evolved fenestrae in their skulls independently of one another. These openings were bounded by arches of bone that reminded early paleontologists of the apses of Roman buildings, so the suffix *-apsid* (from the Latin word for "arch") was coined to refer, with the addition of descriptive prefixes, to the various arrangements of fenestrae and arches. Because the skulls of the earliest amniotes were archless, the term *anapsid* ("without arch") came to refer to these forms.

Certain descendants of anapsids evolved a single arch bounding a fenestra located high on the skull; because this arch was wide, the term *euryapsid* ("broad-arched") was coined for animals sporting such. Yet another amniote group evolved two fenestrae, one above the other, a condition called *diapsid* ("two-arched"). From these descended our modern reptiles, with the exception of turtles, which were an independent offshoot of the anapsid line; dinosaurs and their relatives also possessed diapsid skulls.

Probably nearly simultaneously with the appearance of the above group, at the very beginning of amniote history, there appeared a group of animals possessing one arch along the lower edge of the skull—a "cheekbone"—with a single fenestra above. Because it was originally (and incorrectly) believed that the two arches of diapsid ancestors had fused to form the single lower arch bounding that fenestra, this group of animals was named Synapsida ("Those with Fused Arches"). Like the Indians of North America, our remote ancestors were misnamed by bumbling neophytes. And while we now know that the "synapsid" condition evolved independently from, and probably earlier than, the diapsid one, the name persists in the great amniote subclass Synapsida from which we mammals descend.

The earliest synapsids form an order Pelycosauria ("Bowl-lizards"), named for their unique pelvic structure, a characteristic predisposing them to fairly quick movements on land. The earliest representatives of order Pelycosauria show a tendency toward use of the higher operating temperatures and greater appetites to which the amniotes were becoming adapted. This direction toward agility and efficiency in capture of prey is reflected in the increasing length of the jaws of such early, relatively small forms as the half-meter-long (one-and-a-half-foot) *Haptodon* ("Fastened-teeth") and its meter-long cousin *Varanosaurus* ("Monitor-lizard," from the modern Varanidae, the living monitor lizards).

The heads of these animals were deep, long, wedge-shaped traps composed mostly of sharp-toothed jaws adapted to hold active, armored prey. While earlier predators had short, cone-shaped pegs for teeth, suitable only for holding rather sluggish food until it might be gulped alive, the early pelycosaurs possessed longer, sharper teeth of differing sizes to provide better purchase on the wiggly insects. The better to pin down their agile food, pelycosaurs came equipped with several good long teeth near the front of the jaws; these longer

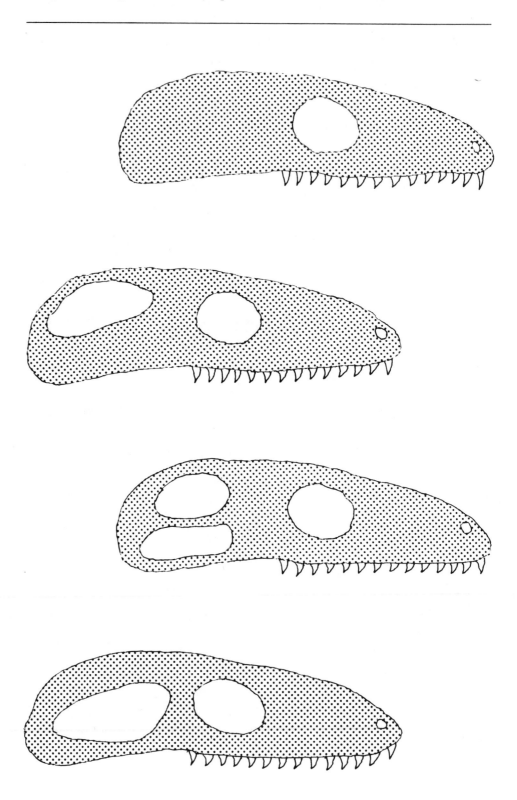

The "apses" of amniote animals. A. Anapsid ("archless"). B. Euryapsid ("broad-arched"), with the opening high on the skull. C. Diapsid ("two-arched"), the form present in living reptiles, and in archosaurs and birds. D. Synapsid, the incorrectly named "fused-arch" style characteristic of our own evolutionary line. Because of temporal structure, the pelycosaur-therapsid assemblage is called the Synapsida.

teeth were the precursors of the canine ("dog") teeth that have remained hallmarks of our synapsid-mammal line through 300 million years of evolution.

These new long jaws and jagged tooth rows were tremendously successful in the warm, sunny Pennsylvanian period, and the early predatory pelycosaur stock quickly diversified to fill a number of ecologic niches based not only on active land predation but also on fishing and other swampy pursuits as the benefits of the newfound jaw efficiency were put to use. This adaptive radiation produced some astounding creatures, the like of which is not to be found alive anywhere today; in fact, they look to us, their descendants, rather like products of some evolutionary disorder.

From some relative of *Varanosaurus* descended a suborder of predatory pelycosaurs, Ophiacodontia, whose name means "Snake-teeth." Reaching their greatest diversity of species during the early Permian period of about 265 million years ago, ophiacodonts seem to have taken to eating slow amphibians and fish in swamps; this probable swamp-loving life-style may account for the general conservatism of the suborder. Life is easy in swamps.

Early in pelycosaur evolution appeared another suborder, Edaphosauria ("Ship-lizards"), whose name is taken from the tendency of some edaphosaurs to evolve a tall fin, or "sail." Edaphosaurs are represented by three

---

The many-toothed lower jaw of *Edaphosaurus*, which enabled it to grind up vegetation. The evolution of plant-eating was a pioneering event that permitted vertebrates to utilize dry-land vegetation and thus fuel a rapid evolution of large land animals.

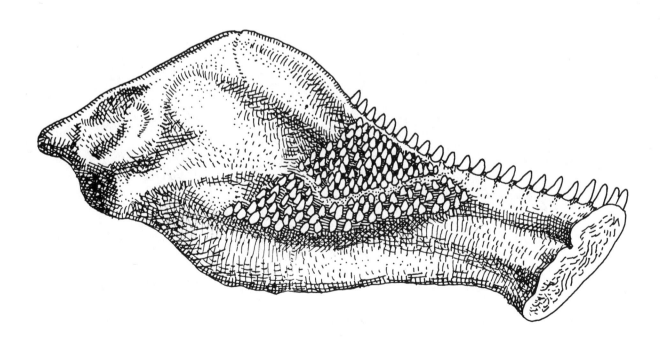

families whose builds and dentition suggest a growing tendency toward herbivory (vegetarianism), a momentous accomplishment in the history of the vertebrates. In tapping the previously untouched potential of plant-eating, the edaphosaurs founded entire ecosystems of herbivores and carnivores of larger size than had heretofore existed on land.

Herbivory permits animals to make more direct use of the sun's energy by eating the energy-trapping plants themselves rather than going through middlemen such as insects. In addition, because plants do not run around, herbivores may be relatively large, slow animals, at least as long as they are not overrun and destroyed by predators. During the rise of the edaphosaurs, of course, there were few landgoing predators larger than a big rat, so no obstacle arose to thwart the appearance of a host of these great (four- to five-meter-, or sixteen- to twenty-foot-long), sausage-shaped "Ship-lizards."

The transition to plant-eating required important metabolic changes in its vertebrate pioneers. Before the herbivores made their appearance, ingestion of insects was the main means by which plant energy was made available to terrestrial vertebrates: insects, having concentrated a good deal of energy in their crunchy little bodies, were easily digestible and easy to find. However, plants alone contain far less energy and nutriment per unit volume than

Haptodon, an insect-eating pelycosaur. From this or an animal very like it, evolved the therapsids, which replaced the Pelycosauria and populated the earth with its first balanced vertebrate land fauna. *Haptodon* may not have been beautiful, but in its chase after the burgeoning insects of the Lower Permian, it represented one of the most advanced expressions of terrestrial life to date.

animal prey does, and herbivores are thus forced to consume more than insect-eaters must in order to keep up their body processes. This results in enlarged digestive systems in most herbivores, meaning that they must increase their total body volume, another impetus toward large size.

However, as we have already seen, increased body volume means a correspondingly smaller body surface through which heat may be transmitted to maintain a constant body temperature (as size increases, volume increases by cubic exponent, power of 3, while surface increases only to a power of 2). Thus, a very large ectothermic ("heated from without," remember?) animal is subject to all sorts of trouble because its means of temperature regulation is so inefficient. One way for a large ectotherm to handle the problem is to remain near water, entering or leaving this comfortably stable medium as its body temperature requires. Modern crocodilians, which are at least partly ectothermic, stick to this mode of existence. During the late Pennsylvanian and early Permian, a great number of immense, obese edaphosaurs did the same.

Another way to solve the temperature problems caused by large size is to increase the body surface–to–volume ratio to a point where the animal is again able to handle temperature transfer efficiently, and certain edaphosaurs did precisely that through the evolution of the famous "sail" for which they are named. This structure consisted of a skin membrane supported by elaborate spines rising from the top of the animal's vertebrae, across which a rich network of blood vessels carried its blood. When oriented across the sun's rays, the sail collected a great deal of solar heat, which was transmitted by the

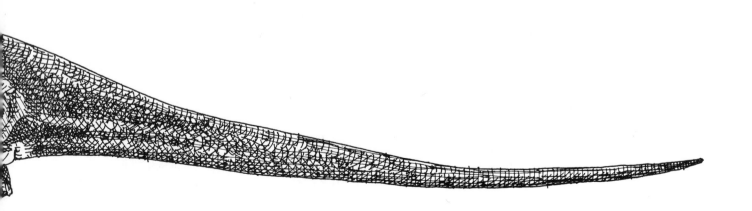

A Lower Permian bog nightmare, in which *Edaphosaurus*, the large herbivorous model to the right, is attacked by *Dimetrodon*, a highly predatory version of the same principle: surface area increases only by squaring, while volume increases by cubing. Although the "sails" on these animals resemble one another in both form and function, they evolved in parallel rather than from a sail in a common ancestor, in response to the need for a larger body surface in both herbivore and predator. *Dimetrodon* and *Edaphosaurus* stayed close to swampy lowlands, but smaller sphenacodonts like the little fellow at the lower right were to found a new dynasty of upland animals, the therapsids.

The skull of *Dimetrodon* illustrates the origin of canine teeth and the snaggly tooth row from which the mammalian chewing apparatus would ultimately evolve.

blood to the rest of the animal's body in a relatively short time; thus in the morning the big edaphosaurs might heat right up and go to work before their little predatory cousins could catch them. Conversely, if an edaphosaur felt that it was overheating in the noon sun, it could turn itself parallel to the sun's rays so that both sides of the sail became radiators, passing excess heat from the blood to the surrounding air. This convenient arrangement permitted edaphosaurs to become one of the longest-lived lines of pelycosaurs.

The appearance of the edaphosaurs as a vanguard of land-dwelling vertebrate herbivores might be expected to be followed closely by an adaptive radiation of some form of predator large enough to eat such weighty animals. This was indeed the case. While they had never completely abandoned the long, fishlike body plan and low-bellied posture of their more primitive cousins, many insectivorous pelycosaurs had experienced a slow tendency to raise their bodies more and more off the ground while walking. This was accomplished through a slimming of the bones of the legs, which were equipped with powerful muscles to elevate the body off the ground for short dashes after insects in what I like to call the "half-pushup" position. To test this limb arrangement, you should kneel on the floor, placing your hands palm downward in front of you, and bend your elbows, allowing your body and face to come within three or four inches of the floor. You are now touching the floor only with your knees and hands, and your elbows are bent so that your upper arms are parallel to the floor and your lower arms are perpendicular to it. Try holding this position for a while, and you will see that pelycosaurs had a fairly rough time getting around, especially if they experimented with any increase in weight.

Nonetheless, selective pressure among small insectivorous pelycosaurs forced some to experiment with tapping the large amount of energy stored in big, fat edaphosaurs. The process probably began in some form rather like *Haptodon* and gave rise to a third suborder of pelycosaurs, the Sphenacodontia ("Wedge-teeth"), which produced a variety of predators, many of which were more than able to overcome the largest of edaphosaurs. Sphenacodonts, appearing in the Upper Pennsylvanian about 270 million years ago, ultimately became the most advanced pelycosaurs both in limb structure and in tooth arrangement, and served as a route through which more sophisticated synapsids would ultimately evolve.

Early sphenacodonts were pretty much like *Haptodon* in appearance and insect diet. However, their fossil record early shows a tendency to chase after the large herbivores of their time; in conjunction with this tendency, their jaws became deeper to support the action of larger and more powerful muscles, while their canine teeth grew into long sabers with which they could open up the corpulent, sluglike edaphosaurs as if they were cans of beans. Forward of the canine teeth in advanced sphenacodonts, a set of short, sharp incisor (cutting) teeth permitted these dimly ferocious animals to nip pieces of tough meat from their prey, while to the rear of the canines a serrated set of short teeth, the forerunners of molars in more advanced animals, enabled them to chop their meals into neat pieces for efficient swallowing. Sphenacodonts were among the first terrestrial predators to give up swallowing their prey whole, a significant advance in food-processing technique that would someday give rise to true chewing. But not yet.

Parallel evolution is the process in which different organisms experiencing similar selective pressures tend to evolve similar structures—rather as, in a modern technological event, both the United States and the Soviet Union produced missile-launching submarines in response to selective pressure exerted by paranoia. In an interesting example of parallel evolution in the Lower Permian, certain higher sphenacodonts began to evolve "sails" along their backs to increase their control over body temperature, just as the

38

B.

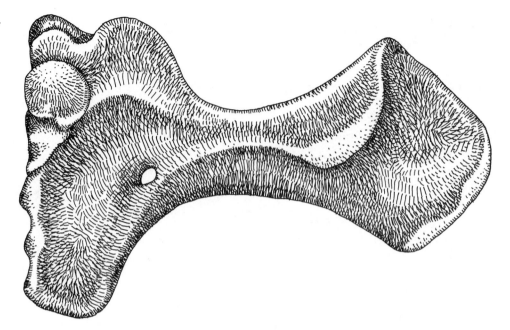

A.

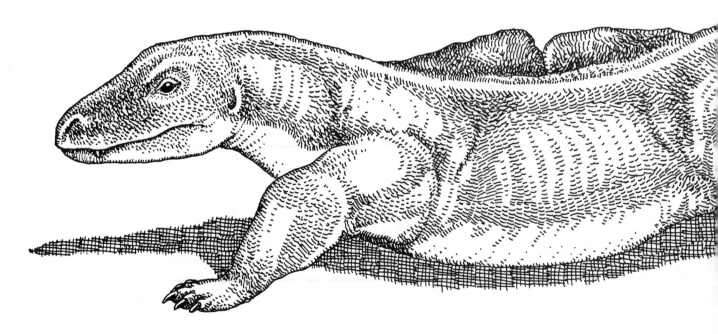

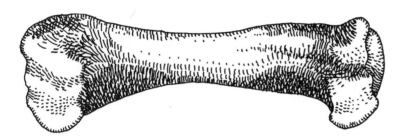

A. The primitive pelycosaur *Varanosaurus*, illustrating the "half-pushup" posture characteristic of the group. B. The humerus (long bone of the upper forelimb) of a pelycosaur reflects in its contorted form the tremendous stresses to which it was subjected by the "half-pushup" gait; compare this bone to the humerus of a fully erect mammal of similar mass (C).

C.

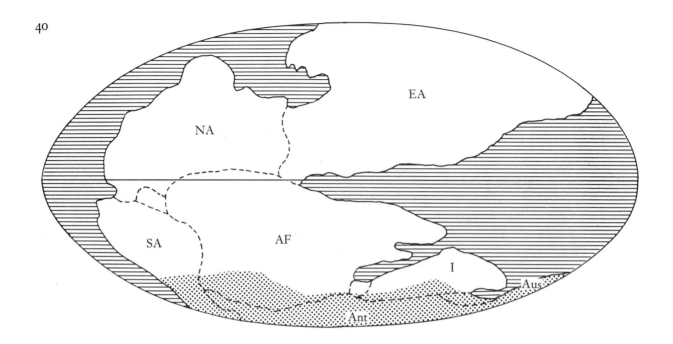

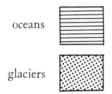

oceans

glaciers

The Pangaean landmass during the early Permian, with glaciers covering the south of the landmass. Therapsids evolved from pelycosaurs in response to pressures from the uncertainty of the glacial times. About 260 million years later, another stressful glacial age would see the founding of humanity among remote therapsid descendants. Every cloud has a silver lining.

edaphosaurs had before them. Sails allowed big sphenacodonts to regulate their exchange of heat with the environment at least as well as edaphosaurs could, and thus to catch and consume said edaphosaurs. Perhaps best known of the sphenacodonts is the "fin-backed" *Dimetrodon* ("Two-measure-teeth"), one of the commonest animals in the early Permian and a great evolutionary success for ten million years or so.

We can imagine what early Permian life must have been like for the pelycosaurs, which were for their time the most advanced forms of terrestrial life. While the little ones chased lizardlike after cockroaches and other improbable insects, the big ones lay around on their bellies resting their short legs until some great physiological need—for food, water, sex, or a change of temperature—overcame them. At such times they'd set off in pursuit of same with their bellies perhaps lifted a short distance off the earth by their short but sturdy legs. While this was an inefficient gait at best, it functioned well enough when the animals were warm and energetic, thus reflecting a way of life dependent on fairly high body temperature soaked from the sun and, in the

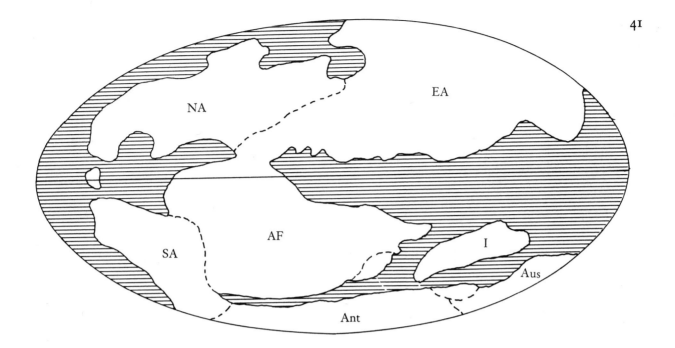

The breakup of Pangaea during the Upper Triassic. During this time of the scattering of the continents, many wonderful organisms were appearing. Notable among these were conifers and flowering plants, new insects to feed upon these, and early dinosaurs and mammals. Abbreviations: NA = North America; SA = South America; EA = Eurasia; AF = Africa; I = India; Aus = Australia; Ant = Antarctica.

case of predatory pelycosaurs, fueled by the presence of a variety of slower animals from which to select their meals.

Thus, across the incredible landscape of the Permian around 260 million years ago, these wonderful sailed and waddling creatures grazed, hunted (and were hunted), made love, and died by the millions, generation after generation, refining further and further the physiology of terrestrial living and strenuous activity. As all times are in a living world, these were times of change; while the large edaphosaurs and their vegetarian relatives reached an equilibrium for a time with the big sphenacodonts that preyed on them, a variety of smaller sphenacodonts experienced selective pressure from each other and their bigger cousins, and from increased speed in insects, to move faster and use energy more efficiently. It was among these little fellows that the changes of the first half of the Permian were to work their greatest wonders, to found the next great adaptive radiation of synapsids.

The early Permian was a period of global revolution 35 million years long. Vast changes in climate, geology, and biology dictated patterns of evolution that have reverberated through the 230 million intervening years to modern times. The continents of the Permian were largely incorporated in a single gigantic landmass, roughly hourglass-shaped, called by modern geologists Pangaea ("All-land"). Of this hourglass, the northerly half was composed of what would one day be North America, Europe, and parts of Asia, while the vast southern continental mass comprised what are now called Antarctica, southern Africa, South America, India, Australia, and parts of southeastern Asia.

Across the warmer, equatorial areas of Pangaea, through North America

---

Caseid pelycosaurs, which were among the first large vertebrates to stray far from the swamps. The caseids were vegetarians, and lived in places where there were as yet no predators to bother them; hence, because they neither had to chase their food nor flee their enemies, caseids accurately reflect in their appearance the effects of uninterrupted easy living. Compare them to the condition of the culture of the United States at its mid-twentieth-century height.

and Europe, pelycosaurs represented the apex of evolutionary achievement among the waddling vertebrates of the times. Limited by their need of warmth, larger tetrapods were unable to wander too far into the cool south lest the cold winter overtake and destroy them. Most such animals represented evolutionary dead ends, their futures restricted by their highly specialized *modi operandi*. Only the smallest land vertebrates, those able to hide themselves in the mud to sleep out the winter, could spread out at all.

In those remote times, at the height of pelycosaurian evolution, the world was experiencing a cooling trend comparable to the far more recent periods of glaciation that we call the Ice Age. Especially around the South Pole, which was then, as now, located somewhere in Antarctica, the cooling caused a gradual buildup of snow and ice that started a flow of glaciers, vast rivers of ice several kilometers thick, that slowly spread across much of southern Pangaea. At the edge of these expanding ice fields lay a tundra, a cold desert sparsely populated by cold-adapted vegetation that remains largely unknown to us because the tundra environment is not suited to fossilization.

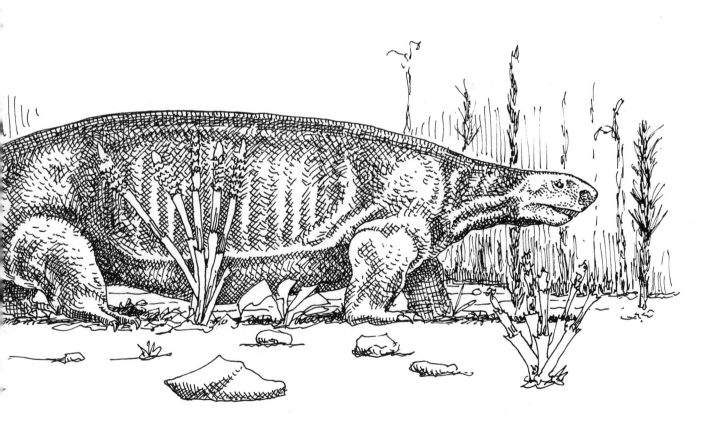

Farther from the edge of the ice, a genus of seed fern called *Glossopteris* ("Tongue-wing") dominated the vegetable world. Seed ferns were primitive plants whose spores were fertilized in moist conditions, and the long, cold rainy season of temperate Pangaea perfectly suited the forests of tree-sized seed ferns. Such forests remain today in the form of vast coal deposits, and their winglike seeds are among the commonest Permian fossils. The *Glossopteris*-dominated flora suggests a climate similar to that of the American North Woods today; all large ectothermic animals such as pelycosaurs would have been excluded from such an environment because of their imperfect internal heating systems.

Nonetheless, the pelycosaurs were a very successful and progressive group of animals, and they filled their appointed equatorial wet-lowland econiches quite tightly. During warm periods in the Permian climate, these comparatively mobile animals extended their ranges, the herbivores following warm-weather vegetation south, and the predators following the herbivores. When the weather turned cold again, the pelycosaur range was once more compressed toward the north and the tropics. This cycle occurred over and over again during the Lower and Middle Permian, exerting tremendous selective pressure on all living things to adapt.

Periodic locking of large quantities of the earth's water into glaciers exerted more pressure on pelycosaurs by causing arid periods during which the size of their habitats was reduced through lack of water. The small ones, especially, suffered as reduction in space and resources forced them to leave the lowlands more and more frequently in search of prey. As they did so, selective pressure increased on them to improve their motility—they were forced not only to hunt ever faster insects but also to stay out of the way of their own brethren. Like their remote descendants, human beings, pelycosaurs tended to be hard on one another in tight situations; living in the upland Permian was as hard for them as living in the oil-tight United States is for us.

Some plant-eating pelycosaurs responded to the changeability of the Lower Permian climate by invading upland forests, leading a vegetarian vanguard to these regions. Here they found no competition, for the only upland grazers before their time were various insects related to our modern grasshoppers. Prominent among these new dry-forest pelycosaurs were members of a family Caseidae of long, ponderous creatures shaped rather like rolling pins.

Their ridiculously small heads and vast, cylindrical, short-legged bodies (up to four meters long—about thirteen feet—quite large for early Permian land-dwellers) suggest that caseids lived an easy, leaf-munching, predator-free life. They were far too big for their insect-chasing cousins to bother, and no

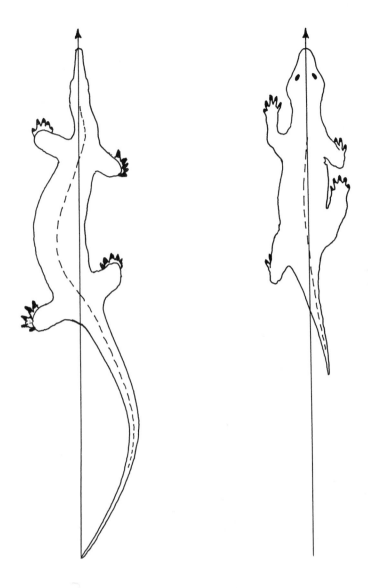

Taking a walk. *Left*: the pelycosaurian waddle. *Right*: the more rigid body of the therapsid permitted its legs to lift its body off the ground by rotating somewhat underneath. This improvement increased energy consumption, an increase reflected in the growing efficiency of the therapsid jaws.

large predators such as *Dimetrodon* had left the Lower Permian swamps in pursuit. In fact, the caseids give us an unequaled look at what uninterrupted easy living can do to one; their sluglike builds and witless lives were the logical outcome of an existence free from strife of any sort. Thus, during much of the first half of the Permian, tremendous amounts of upland plant energy was turned into caseids, a process uninhibited by the impolite intrusion of any meat-eaters capable of harming them.

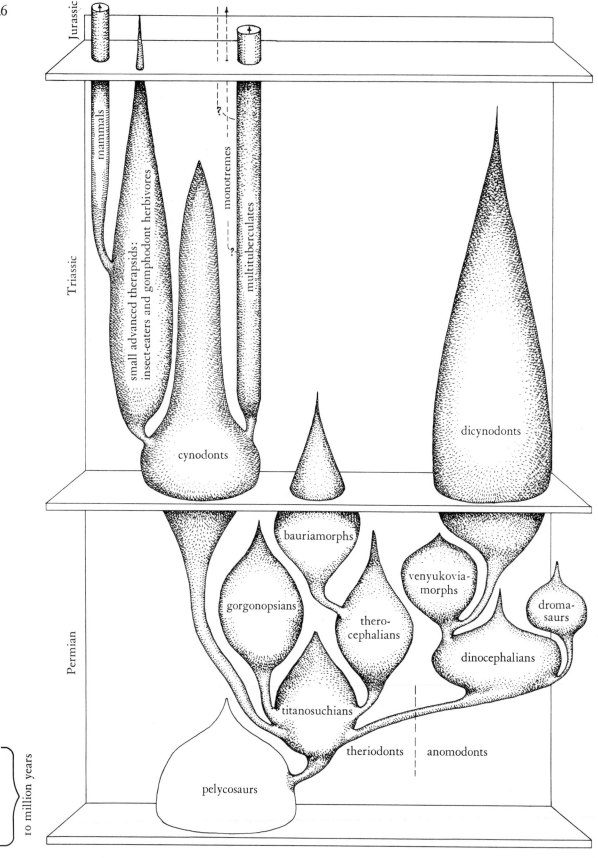

Jurassic

Triassic

Permian

mammals

small advanced therapsids:
insect-eaters and gomphodont herbivores

monotremes

multituberculates

cynodonts

bauriamorphs

gorgonopsians

thero-
cephalians

venyukovia-
morphs

dromo-
saurs

dinocephalians

titanosuchians

dicynodonts

theriodonts   anomodonts

pelycosaurs

10 million years

Pelycosaurs were not the only groups invading the uplands in the form of vegetarians. Other, unrelated reptile groups—notably cotylosaurs of the family Pariesauridae ("Walled-lizards," named for the shape of their skulls)—were experimenting with the same way of life, but were getting better at it. In addition, smaller animals related to the various insect-eating groups then invading the uplands experienced pressure toward vegetarianism as the competition for insect prey increased. So, throughout the Lower Permian, large numbers of primordial herbivores gradually came to inhabit the upland forests of tropical Pangaea in a unique era of predator-free peace the like of which has not since been seen on this planet.

Now, it is the way of life that no one ever finds a perfect existence—someone else always moves in to spoil it all. Therefore we could, had we been around during the early Permian, have warned all these new upland vegetarians that someone, someday, was sure to figure out how to eat them and thus mess up their luxurious existence. But we weren't there. No one warned the vegetarians, and the inevitable finally happened. We're lucky it did, too, for among those who first discovered how to eat those herbivores were our own ancestors.

Pressure toward preying on big upland vegetarians was strongest on the small, insect-eating sphenacodont pelycosaurs who shared their environment. Some of these comparatively fast little animals evidently took to eating smaller vegetarian herbivores such as small caseids. Because the caseids were so plentiful, they represented a food source so rich that the sphenacodonts eating them experienced a tremendous adaptive radiation—but in so doing they ceased being pelycosaurs and took on a new identity.

The chief advantage in being a little sphenacodont was speed. Increase in size could have two effects on speed: either it might cause them to become great, slow gobbets of flesh like their caseid prey (in which case, of course, they'd starve), or it would exert considerable pressure toward a physique retaining the speed advantage, but at great cost in energy. The more you weigh, the more you must eat to lug yourself around at any reasonable velocity.

A therapsid evolutionary tree. The scale (*lower left*) is ten million years long in this drawing. The tremendous Permian adaptive radiation of early therapsids is clearly shown here, as is their gradual Triassic petering-out in response to the rise of the dinosaurs.

Although not a "natural" group in themselves, the small advanced cynodont therapsids from which mammals would ultimately evolve are lumped together in a bunch (*far left*) to emphasize the importance of their small size and secretive ways in founding a mammalian way of life.

A very primitive therapsid, *Phthinosuchus* ("Waning-crocodile"), little removed in appearance from sphenacodont pelycosaurs; nonetheless, its gait was probably rather more efficient and its skull more solidly constructed.

Response to this pressure came with gradual improvement in the stance. When pelycosaurs walked, they carefully picked up one leg at a time, while their bodies swung fishlike in conjunction. When either of the splayed forelegs was lifted to take a step, the shift in the animal's center of gravity would, had it not possessed a long tail as counterweight, have caused it to fall on its face. But the tail was otherwise pretty much useless, a dragging weight that interfered considerably in the competition for speed.

So, evolving toward an ability to eat big herbivores, upland pelycosaurian predators experienced pressure toward larger size, more efficient walking, and better jaw mechanisms with which to handle increases in appetites resulting from their new life-styles. The food was there in plenty, and a new synapsid order exploded across the planet in the greatest adaptive radiation of vertebrates the continent of Pangaea had yet seen. This was the rise of the order

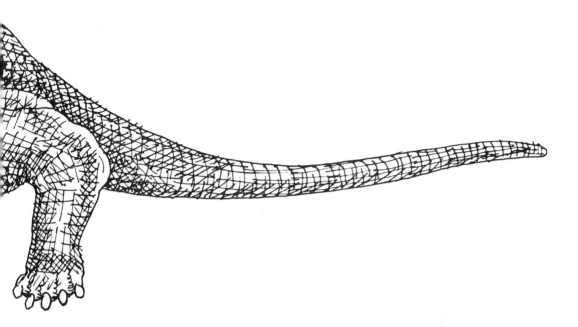

Therapsida, an enormously successful and diverse group of animals and the first to produce a balanced vertebrate ecosystem on land.

The word *therapsid* refers to the characteristic mammallike arch of the cheekbone in this order. The shape of this cheekbone is important because from it hang some of the muscles of the jaws—if we didn't possess these muscles, we'd have to be fed through tubes. Early therapsids, having no tubes, caught all of their prey with their teeth, and bite power was essential to their survival. The shape of their cheekbones suggests that these first therapsids were little improved over their ancestors in this regard, having little more leverage in their jaw mechanism because of inefficiently arranged biting muscles.

You can already guess, then, the significance that cheekbones had in the evolution of our ancestors, let alone their importance in the living higher mammals, such as Katharine Hepburn, the highest exemplar of these to date. Cheek configuration dictates and reflects jaw efficiency, which in turn reflects an animal's ability to capture prey, which in turn reflects that animal's ability to fuel itself, whether it is a lazy ectothermic bum, or a high-stepping endotherm like Katharine Hepburn, or whatever the stage in between these extremes. The study of paleontology is full of surprises.

Because of the ordinarily very gradual pace of evolution, we might expect that the "lowest therapsid" would be an animal little removed physically from its "highest pelycosaur" ancestors, such as those found in North American Permian fossil deposits. Such animals appear in Russian Permian deposits slightly more recent in the fossil record than the American ones, and at this dividing point our typically arbitrary taxonomist buddies have founded the therapsid order atop that of the pelycosaurs.

Primitive therapsids were dog-sized animals with faces very like those of their sphenacodont antecedents, except that the cheekbone arch was longer.

An early titanosuchian therapsid predator of the sort that exploited upland caseids for food and hastened the fall of the Pelycosauria. Such carnivorous therapsids replaced pelycosaurs as the fastest land-dwellers of the Middle Permian world.

They could bite harder. This small advance does suggest a bigger appetite and one more efficiently looked after, and the limb bones of these animals indicate that the extra food was going largely into faster walking (running hadn't been invented yet).

Eating of the big, slow, squashy vegetarians in the uplands quickly resulted in a tendency toward large size in early therapsids. Representative of the breed were the stocky predators of infraorder Titanosuchia ("Titanic crocodiles"). These were dog- to bear-sized animals admirably designed for mushing up their smaller mid-Permian relatives and the remaining pelycosaurs, by means of great, jagged teeth in mighty, crushing jaws. Although it is probably unscientific to say so, these do appear at first glance to have been some of the most astonishingly ugly animals ever to have been produced by the therapsids, or by any other vertebrate line. Nonetheless, we ourselves are probably descended from some such animals.

Titanosuchians may have been among the first therapsids—indeed, the first

vertebrates—to spend much time in cooler climates. Their remains are plentiful in Middle Permian South Africa, and they appear to have been quite comfortable in that cold time and place. Their large size and stocky build permitted titanosuchians easier control of body temperature in such rigorous climates, where smaller size can be a liability because of the comparatively great surface area–to–volume ratio of small animals.

It is among the limb bones of these early therapsids that we find the most significant evidence of their increasing metabolic sophistication. The old "half-pushup" stance of the pelycosaurs had been abandoned by titanosuchians in favor of a "three-quarters-pushup" posture in which their limbs were rotated partly *under* their heavy bodies, rather than off to the sides. Thus, titanosuchians no longer lay about on their bellies so much, but were obliged instead to stand up off the ground, thereby expending much more energy. The new, more erect posture increased walking efficiency by lessening the intensity of the muscular effort required by each step, but it also required constant activity of the muscles about the limb girdles for support. This posture, then, probably served also to raise the temperature of the titanosuchian, further aiding the animal's ability to handle cool climates, for all muscle activity does release heat. Therefore, many titanosuchians were probably able to extend their hunting hours somewhat into the evening, a further advantage over their contemporaries.

Animals able to turn muscle heat to their own advantage in this fashion were thus freed to some extent both from restrictions imposed on their range by temperature variations and from the morning-to-evening (nine-to-five), sun-dependent timetable of their ancestors. Unlike the other vertebrates of their day, therapsids were beginning to liberate themselves from the need for collecting direct solar energy (lying in the sun), by evolving an ability to collect solar energy from their food instead. In addition, they were getting faster and bigger.

It is hardly surprising, then, that the early therapsids experienced an adaptive radiation of unprecedented magnitude, quickly afflicting the Permian world with a host of weird animals that reflected the wide variety of habitats open to these pioneers in active living. As always in the case of a successful adaptation, the descendants of the innovators began to press on one another as they mopped up the resources of their new econiches. They ate up all the fat caseid herbivores, and they ate up the remaining fat edaphosaurs. They ate up the carnivorous pelycosaurs, superseding these as dominant predators in the terrestrial world. Then, inevitably, the titanosuchians began to eat each other up as well.

Such internecine warfare forced them to expand their ranges ever further in search of food and safety from their brethren. It also forced them toward greater walking efficiency and greater metabolic efficiency, and compelled them to carve out a new world of ecologic niches. Titanosuchian descendants remained the dominant vertebrate forms from the Middle Permian through most of the succeeding Triassic period, a reign of some 70 million years. During this time, the therapsid line produced a wide variety of preposterous carnivores and herbivores comparable in their diversity to the multifarious modern mammals. These creatures delightfully populated an outlandish world of strange plants and animals that continues to astound us now, 180 million years after they died off.

ANOMODONTIA

All of the first therapsids were predators, and there ultimately evolved from these primordial meat-eaters an astounding array of therapsid "cats," "dogs," "weasels," "hyenas," and what have you. These varied hunters are loosely grouped in a suborder Theriodontia ("Beast-teeth"), which, however goofy-looking to us in retrospect, were not only the most formidable animals of their time, but were also the founders of our own line. As we all know, meat-eaters must eat meat, and meat is made from plants by vegetarians. Therefore, the evolution of more progressive carnivores had to await the advent of sophisticated herbivores to exploit the wide variety of vegetation available during the Middle Permian.

The coming of plant-eating therapsids was the most important thing to happen to the infant therapsid order. Their comparatively great appetites forced on plant-eating therapsids a complex of chewing and digestive advances that permitted them to eat almost anything vegetable that the Permian had to offer, and their resultant success provided the central impetus for the evolution of the therapsids—and, indeed, the entire terrestrial vertebrate world during the Permian and Triassic periods. Although a ludicrous-looking crew to modern eyes, these remarkable vegetarians fueled the development of the first truly balanced vertebrate faunas to inhabit our planet's continents.

As we have seen, the earliest therapsids belonged to the infraorder Titanosuchia. Named for a genus *Titanosuchus* of dog- to bear-sized, meat-eating waddlers, the group appears to sit close to the division between more advanced therapsid carnivores, on the one hand, and the plant-eating bunch, on the other. As food resources for early carnivorous therapsids declined in response to their successes, some of these animals responded by enlarging their

54

Herbivores that shared the Upper Permian and the Triassic with anomodont therapsids. *Above*: an Upper Permian pariesaur, whose gait parallels that of the anomodonts, reflecting similar selective pressure from identical predators on this holdover from the stem reptiles. *Below*: a rhynchosaur ("beaked lizard"). Rhynchosaurs occurred by the thousands in Triassic landscapes, probably in response to the appearance of early seeds which they cracked with their sharp beaks. Both pariesaurs and rhynchosaurs served as food for carnivorous therapsids in their respective times.

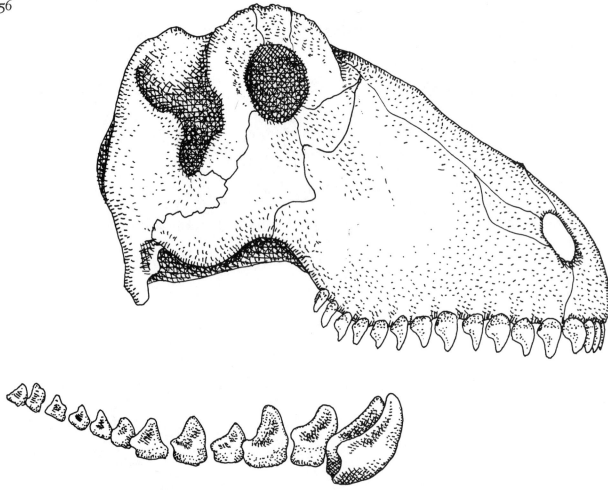

The skull and teeth of *Ulemosaurus*, a large Russian dinocephalian herbivore, showing the tendency of its dentition toward vegetarianism.

diets to include a variety of substances other than neighboring animals. This trend early resulted in the founding of a suborder of largely herbivorous therapsids, which replaced and far outperformed the plant-eaters that had fueled their ancestors. This vegetarian group has been lumped into a suborder Anomodontia, named for the tusks of some that had initially been dubbed "nameless-teeth" because their possessors were as yet undiscovered by the paleontologists who named them. The process in which a progressive group of carnivores (in this case early therapsids) presses some of its descendants into herbivory is a common one in the story of earthly evolution; a variant of that old theme may have produced the birth of agriculture among the ancient hunting human beings.

Although most titanosuchians were squat, powerfully built animals with fearsome saberlike teeth adapted for piercing the tough hides of their contemporaries, one family, the Jonkeriidae of southern Africa, early showed tendencies toward vegetarianism. Their long faces contained, in addition to sharp canine teeth, small grinding cheek teeth adapted more to the crushing of plant matter than to the shearing of flesh. Their likely vegetarian life-style is exemplified in the fossil rémains of the type genus *Jonkeria*, great chubby, zeppelin-shaped animals found in the Middle Permian of South Africa. These were some of the first vegetarian vertebrates ever to have lived in cool, temperate climates.

Other titanosuchian descendants occupied an infraorder Dinocephalia, comprising massive herbivores of herding habit. Dinocephalian means "horrible-headed," a possibly biased but genuinely accurate description from our

A jonkeriid, one of the earliest herbivorous therapsids, in the process of experiencing a bit of selective pressure from two of its titanosuchian cousins.

Estemennosuchus, whose head was modified as a weapon of defense. You can see why they named these animals Dinocephalia—"Horrible-headed."

point of view. The teeth of dinocephalians were adapted toward a vegetation-biting and -crushing mode of life that is further reflected in their rotund bodies. In order to utilize the great quantities of vegetation required to sustain their bulk, dinocephalians must have possessed relatively huge digestive systems, for plant matter is far less nourishing per unit volume than the animal food of their ancestors. The great dinocephalians, many weighing more than a thousand kilograms (about one ton), represented a vanguard that, meeting no competition, diversified into a dynasty of funny-looking but successful herbivores that were to outlive most other therapsids and hold sway among our planet's vegetarians until the rise of the dinosaurs millions of years later.

Dinocephalians were the first anomodonts, and are typified by *Estemennosuchus*, a large Middle Permian herbivore that lumbered about the cold reaches of what was one day to be Russia. These animals weighed in at about seven hundred kilograms and were about three meters long; they moved in herds and apparently cropped plants that were lower than their heads, for their faces were sharply downturned from the rest of their skulls in the manner of modern grazing mammals. As might be expected from their large size, dinocephalians were corpulent animals; nonetheless, their semierect posture suggests that they were capable of moving very quickly if excited, a reflection of considerable selective pressure from the therapsid predators of their times.

The dinocephalians were extremely successful, producing by the time of the

---

Skull and restored head of *Venyukovia*, a form transitional between dinocephalians and dicynodonts on the scale of therapsid vegetarians.

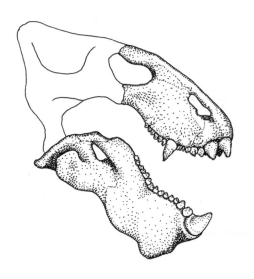
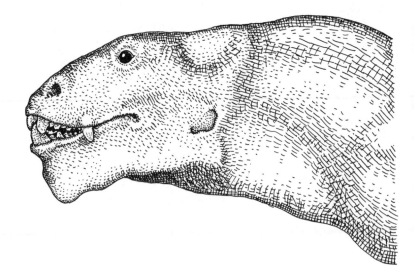

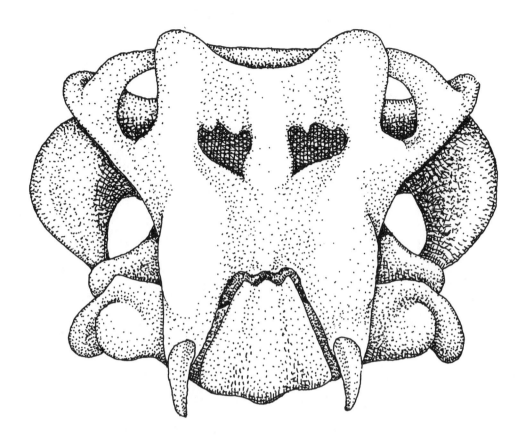

Head-on view of a dicynodont skull, showing the relationship between the cutting edge of the upper jaw and the hooked tip of the lower jaw.

Middle Permian about twenty-five genera of efficient browsers and grazers of various sizes. Some of these, such as *Tapinocephalus* from the Middle Permian of South Africa, possessed heavy bone ornaments and bosses on their heads, suggesting that they were given to combat, either within their own species or with large carnivores. If the latter is the case, it is possible that these animals had a fair degree of social or herd consciousness and that they protected their own from attack.

The immense skulls and thick bones of dinocephalians indicate that they were powerful animals, many of them able to eat such tough and (to me) unappetizing materials as the trunks of seed ferns and the leaves of early conifers and cycads (primitive palmlike plants). Their ability to do so permitted them a phenomenal adaptive radiation in the Middle Permian, one result of which was the origin of the great infraorder Dicynodontia ("Two-dog-teeth") of herbivores that were to wander in immense herds throughout the

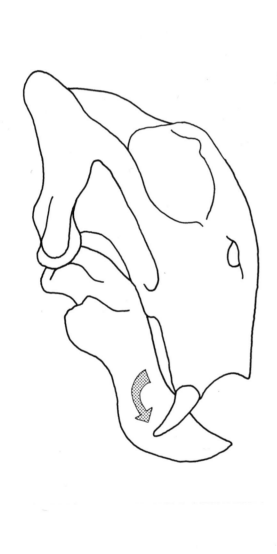
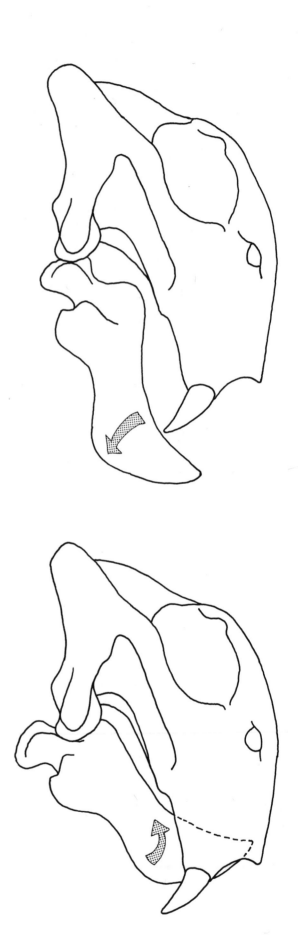

Upper Permian and during most of the Triassic across the supercontinent Gondwanaland, the southern part of Pangaea. Dicynodonts descend from dinocephalian herbivores through a family Venyukoviidae of Russian anomodonts little removed from their dinocephalian cousins in appearance. Venyukoviids possessed huge canines, but their other teeth were considerably smaller, a condition transitional to that of the dicynodonts, in which vegetation was very efficiently ground up in toothless, or nearly toothless, "jaw-mills."

Dicynodonts came in all sizes and several habitats, from the high plains to the swampy bottomlands, where they appear to have lived rather as hippopotamuses do today. In all of them, the skull was sharply modified for the eating of heavy, coarse vegetation of some sort. The length of the head was mostly composed of attachment for the mighty muscles that operated the hooklike jaws, while the snout was modified into a beak (rather like that of turtles), with which dicynodonts appear to have been able to bite into such things as tree trunks, which more discerning animals left alone.

Whatever it was, the dicynodont diet fueled millions of these beasts in herds so huge as to cover the land in some areas. Crowds of them also mired themselves in mud (apparently they were not particularly wary), and their bones were often fossilized in the standing position in which the animals perished. In some forms, such as the highly specialized *Lystrosaurus*, both males and females possessed tusks, whose tips show considerable wear, indicating that some of these animals may have rooted for food as swine do now. In those forms where only males possessed tusks, some sort of ritual combat may have occurred among them for territory as among deer and certain sheep today. This in turn suggests a fair level of social integration based on some sort of pecking order and harem selection among the males of genera so equipped. In some ways, life hasn't changed much since.

Although most dicynodonts were rather large and heavy, they came in small sizes, too, weighing only a few kilograms each. Their complex chewing mills enabled them to profit from all sorts of vegetarian econiches and to outlive most other therapsids. Ultimately, dicynodonts were replaced in the late Triassic period by herbivorous dinosaurs; coincidental with their gradual disappearance was the decline of seed ferns and other primitive land plants on

---

The unique workings of the dicynodont "jaw mill" permitted these vegetarians to enjoy a great Permian adaptive radiation and to survive far into the Triassic after most other therapsids had become extinct. *Top*: the jaw opens; *middle*: it slides forward and hooks into some fibrous vegetable mass; *bottom*: the lower jaw slides backward into the mouth, snapping the vegetable fibers against the tusks grinding them against the roof of the mouth.

Below (*left*): a small dicynodont of genus *Dicynodon*, illustrating the specialized chewing equipment with which its kind achieved their great success. *Below* (*right*): a member of the dicynodont genus *Lystrosaurus*, which inhabited temperate Triassic Antarctica. In life, this specimen probably weighed about as much as a horse. *Opposite*: the skulls of different species in one genus (*Lystrosaurus*) of dicynodonts, showing a variety of adaptations to different vegetable diets.

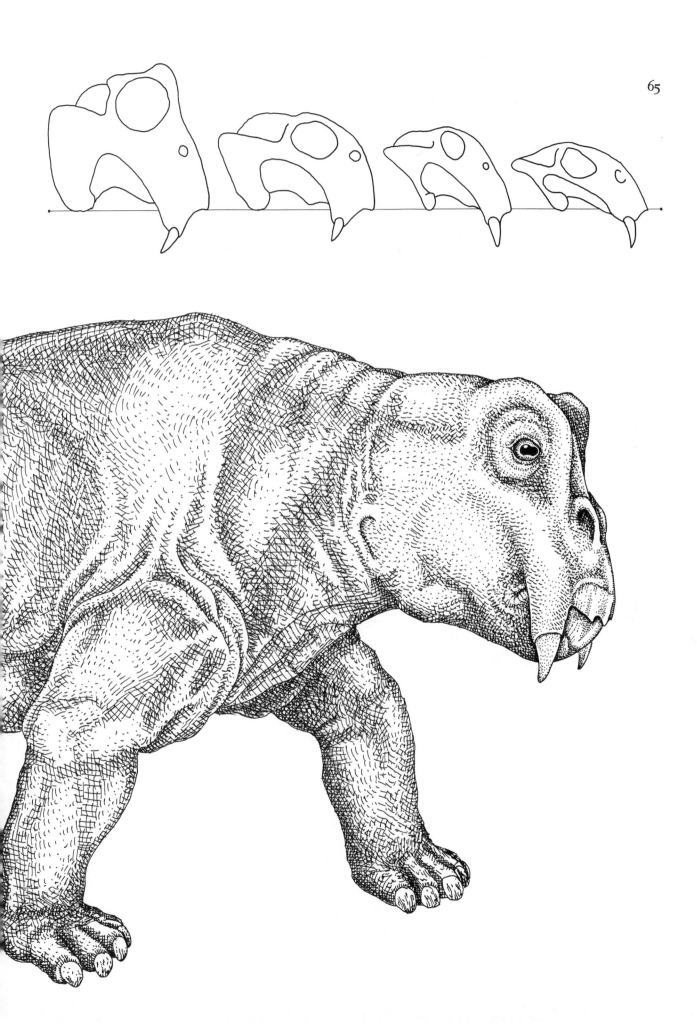

Early dicynodonts retained some teeth. Restoration of *Endothiodon*, a primitive dicynodont possessing cheek teeth but lacking tusks; and its skull.

Emydops, a woodchuck-sized dicynodont in which a few teeth remained as part of the cutting mechanism, and its skull.

which they lived, and the rise of the cycads, conifers, and other, more advanced plants characteristic of the Mesozoic era and modern times.

A curious offshoot of the anomodont line produced a few genera of herbivores less than a meter long that are included in a separate anomodont infraorder Dromasauria ("Running-lizards"). Dromasaurs were characterized by a comparatively light, agile build, and an extraordinary tail, itself as long as the rest of the animal. In three fossil specimens of dromasaurs the tails are curled around the rest of the animals' bodies in a manner suggesting that they wrapped themselves in their lovely appendages to conserve body heat, as do many squirrels, wood rats, and other long-tailed mammals today.

As vegetarians, dromasaurs were not compelled to chase their food; consequently, their agile structure reflects pressure from equally quick predators. Dromasaurs had very large eyes with which they watched out for these predators; the fact that they do not appear to have made it past the Upper Permian suggests that they did not watch carefully enough.

A dromasaur, whose large eyes and small, agile build suggest that its world was full of swift, dangerous predators.

BRINGING HOME THE PERMIAN BACON

Remember the titanosuchians? They were the ugly pioneer meat-eating therapsids from which the anomodonts evolved. During the Middle Permian, titanosuchians were the meanest things afoot, so ferocious that they pretty much ate each other into extinction. In addition to founding many later therapsid strains, the titanosuchians produced a number of fantastic groups in their own right, such as the jonkeriids we have already encountered.

Another titanosuchian family, the Anteosauridae, included a number of huge carnivorous animals whose faces were variously decorated with lumps and bosses of bone, sometimes resembling eyebrows, perhaps used for head-to-head combat between members of the same species. Such combat occurs when its participants have become successful enough to place stress on available food, space, or other necessities, and has changed little across the intervening millions of years. The anteosaurs appear to have been among the earliest predatory vertebrates to achieve such success, but modern mammal populations such as those human beings inhabiting the United States and the Soviet Union have progressed little beyond this kind of bludgeoning in their settlement of territorial disputes.

The meat-eating titanosuchians were large and fierce, but they lacked what we would call grace or agility. Their rather long tails in some cases still balanced them in their splaylegged walking, and they were thus subjected to twin selective pressures: starvation, as their vegetarian cousins improved on their own gaits and on defending one another in herds, and predation, as other titanosuchians turned to cannibalism. The combined effect was pressure on the titanosuchians toward greater efficiency—higher style, as it were—in the catching and eating of prey. The result was the founding, during the

Middle Permian, of two parallel groups of therapsid predators descended from and superseding the titanosuchians.

Parallel evolution is a commonplace in the history of life. As we have seen, the process occurs when organisms that are not closely related but share similar life-styles or econiches tend to acquire like characteristics in response to the necessities imposed by their environments. Returning to our old evolutionary model, the crossbow, we find that as a projectile launcher carried by a single man, it parallels in many ways the later hand-held firearms called rifles; both sorts of weapon possess very similar stocks, for instance, and their triggering mechanisms are identically operated, no doubt in response to the similar states of mind of their designers and operators.

Among early carnivorous therapsids, such parallelism occurred between the infraorders Gorgonopsia and Therocephalia, both of which superseded the titanosuchians by sharing characteristics of gait that made them the most advanced predators of their time, the Middle and Upper Permian.

Anteosaurus, a large carnivorous titanosuchian therapsid. Members of this genus may have banged their eyebrows together in social combat, rather than slaying one another with nuclear weapons as some of their remote cousins have been known to do.

Gorgonopsians, whose name means something like "terrible-looking" (the Gorgons were three sisters of Greek mythology who were so ugly that the sight of one could turn you to stone), were among the first earthly predators to be able to trot. Their bodies were shorter and more rigid on the long axis than those of the titanosuchians, and were supported by comparatively long legs, whose strides enabled gorgonopsians to catch fast dinocephalians and other tasties. Their gait permitted gorgonopsians to do quite well, and they came in sizes appropriate to catching everything from the larger insects to great big anomodonts.

Therocephalians ("Beast-heads") were mostly large predators of dog- to bear-size sharing the walking improvements possessed by gorgonopsians. They were flat-headed animals, massive in build, typified by such awful-looking creatures as *Lycosuchus* ("Wolf-crocodile"), whose saberlike canine teeth sometimes occurred in doublets like massive forks. A few therocephs even appear to have been among the first vertebrates to experiment with a poisonous bite; their skulls contain chambers for venom glands, and their sharp canine teeth were ridged to serve as hypodermic needles, conducting the poison into their prey.

Therocephalians, gorgonopsians, and their ancestors gulped down their

Scymnognathus, a fierce gorgonopsian predator of the type that replaced titanosuchians during the Middle Permian. Compare this form with the titanosuchian illustrated on pages 50–51 to see its progressive nature (more efficient gait).

food all at once, with no chewing involved. Such animals were content to sit around for days or weeks after a meal while their digestive innards slowly dissolved the great chunks of whatever it was they had eaten (usually their relatives). However, several smaller representatives of each group experienced pressure toward increases in energy use. This came about partly because they had to outrun their bigger relatives, and partly because they were competing

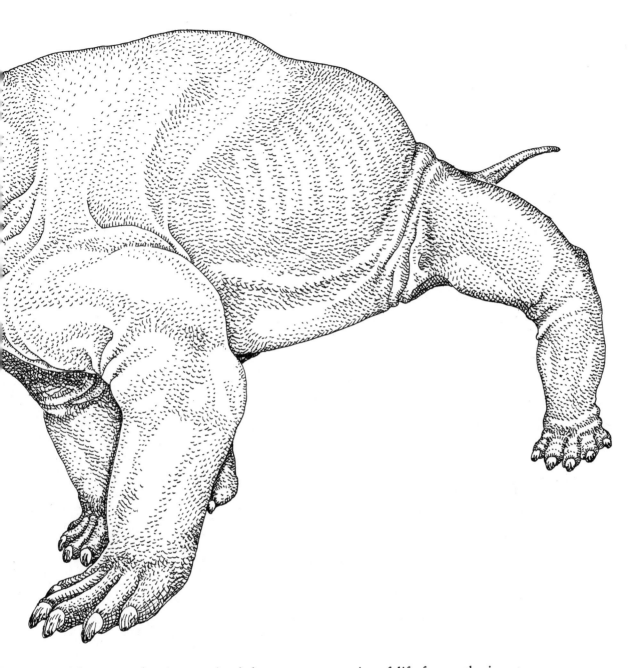

with one another in pursuit of that most progressive of life forms, the insect. As happens so often in our story, insect adaptability was once again the key to a synapsid revolution.

Now, most of us have been admonished by our parents not to talk with our mouths full. While the primary reason for this prohibition is probably the abominable noise that results from the passage of air through partially masticated food, the basic incompatibility between chewing and speaking reflects a more fundamental problem: the close relationship between vertebrate mouthparts and breathing tubes—remember those filter-feeding, gilled ancestors of ours?

In such animals as gorgonopsians and therocephalians, the air breathed in through the nostrils passed right into the front of the mouth and thence to the windpipe at the back of the throat. Being rather phlegmatic metabolically, these nightmarish beasts shared with their ancestors an ability to cease breathing for long periods while they worked immense lumps of food down their throats. During the rest of the time, they breathed, in essence, through their mouths.

A therocephalian predator, *Lycosuchus*, parallel to *Scymnognathus* during the Upper Permian.

Moreover, our insect-eating variety of gorgonopsians and therocephalians were little animals, and because their body volume was small compared with the size of their body surface, they were in greater danger from temperature changes in the variable climate of the Permian than were their larger relatives. Strong pressure toward finer temperature control led to more expenditure of energy in movement by these little hunters, producing greater release of heat from muscular activity to warm them in times of cold weather.

Besides trotting around, the little insect-eaters began moving their mouths faster: they started to chew their food, to chop it into tiny pieces whose total surface area was much greater than that of any single great gobbet. This devel-

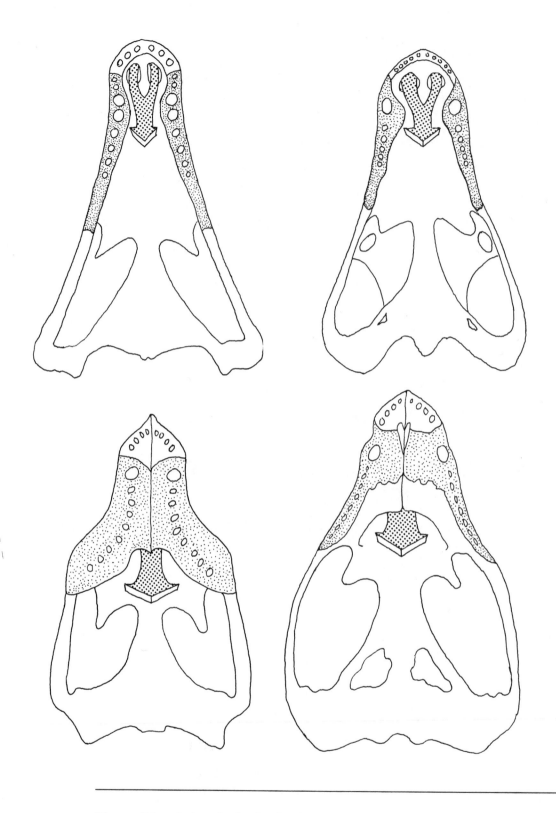

The parallel evolution of a hard palate in two separate therapsid groups. *Left*: a view of the undersides of the skulls of a therocephalian (*top*) and a bauriamorph (*bottom*) descended from it. *Right*: similar views of a gorgonopsian and a cynodont descended from it. In both groups, palates appeared as active living made cutting and chewing of food necessary. Arrows show flow of air; stippling marks maxillary bones from which palates evolved.

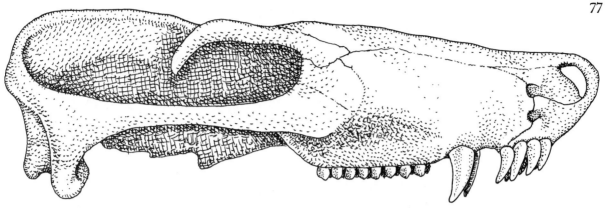

Skull of *Bauria*, a bauriamorph. This is a doglike skull showing the tendency toward a grinding ability that permitted bauriamorphs a high level of activity.

---

opment made for faster digestion; as a result, food energy could be released more quickly into the bloodstream to keep the body temperature fairly constant. Finer temperature control would, in turn, permit these animals to continue chasing their active prey in very cool weather, possibly even during the night. But there was a catch: for this additional energy to be released from food, the little creatures had to take additional oxygen into their bodies to fuel their stepped-up metabolisms.

It was a problem: chewing was necessary for fast energy release; simultaneously, so was faster breathing. In primitive, mouth-breathing therapsids, chewing and breathing could not and need not happen at the same time. In our small insectivorous models, some provision had to be made for simultaneous breathing and chewing, lest they strangle on their dinners. Provision was made, among small therocephalians and small relatives of gorgonopsians, and yet another synapsid revolution took place.

From therocephalian ancestors evolved an infraorder Bauriamorpha (from *Bauria*, a member genus) of lightly built carnivores of rat- to dog-size, which lived in the Upper Permian and Lower Triassic of South Africa. Gorgonopsians or their close relatives spawned the infraorder Cynodontia ("Dog-teeth") of predators of all sizes and, later, even some advanced vegetarians. Both cynodonts and bauriamorphs handled the eating-breathing problem with the evolution of a secondary palate, a plate of bone that diverted the passage of air from the nostrils to an opening back behind the area of the mouth in which food was chewed. The gradual evolution of the secondary palate, so easily visible in the skulls of these parallel groups, reflects more than almost any other characteristic their accelerated metabolisms, their increased dependence on nimbleness, and their large appetites coupled with rapid breathing for enhanced energy release. These innovations are of utmost importance for us because they were to lead to our own way of life. If we were to return to the practice of ancestor worship, a worthy progenitor might lie among these long-gone insect-eating therapsids who pioneered mealtime manners by separating their breathing from their chewing.

The earliest bauriamorphs already showed development of a secondary palate by the time of the late Permian. In addition, they were among the first therapsids to evolve real molar teeth with rough grinding edges in conjunction with the palate. Because they showed a good many advances toward "mammalhood," bauriamorphs have sometimes been held up as mammal ancestors. How-

---

Jaws of a primitive (*above*) and an advanced (*below*) cynodont. Hatching shows the growth of the area of attachment of a masseter muscle permitting a very powerful bite as the energy requirements of cynodonts increased. The circle at the left end of each marks the point of the jaw's hinge against the skull.

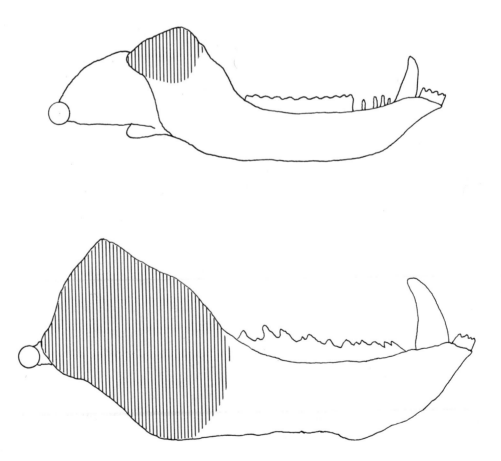

ever, certain specializations among them, plus their early decline and fall, seem to set them well aside from our ancestry.

Cynodonts, on the other hand, experienced a great adaptive radiation in the Middle Triassic; so closely did they approach a mammalian way of life that in some cases we're hard put to tell whether or not certain fossils rate as therapsids or as mammals. It is because of this close approach that the arbitrary jaw-joint boundary was erected by taxonomists to separate the "reptilian" therapsids from "mammals"; otherwise, these two groups are so smoothly connected that at least one paleontologist has suggested reclassifying therapsids with mammals in a great class Theropsida ("Those Who Look Like Beasts"), reestablishing the reptile-mammal boundary between pelycosaurs and the early therapsids. This dispute, however, we will leave to the taxonomists. After all, if there were no room for controversy among taxonomists, there would be unemployment among them. A more dangerous creature than an unemployed taxonomist is hard to imagine.

The parallel approaches to "mammalhood" of bauriamorphs and several cynodont groups reflect a strong Triassic selective pressure on therapsids toward high rates of metabolism and superior endothermy in these originally carnivorous groups, a pressure closely related to the practice of chasing insects. Among cynodonts themselves, the fossil record is so complete, the record of their transition to mammalian ways so clear, that they might be called an "infraorder of missing links."

Among the most primitive of the "Dog-teeth" were members of the family Procynosuchidae ("Before-the-Dog-crocodiles"), carnivorous animals of house-cat size whose secondary palates were as yet incompletely closed across the roofs of their mouths. In such forms as *Procynosuchus*, the family's type genus and an inhabitant of the South African Upper Permian, not only is the hard palate incomplete, but the jaw is still similar to that of gorgonopsians in structure, bearing sharp, cone-shaped "molars" as yet unadapted for anything but piercing (rather than shearing or grinding) prey.

The completion of the procynosuchid secondary palate—and the consequent ability to breathe and chew at the same time—gave rise to a cynodont adaptive radiation at the end of the Permian. At this time there appeared a number of cynodont families; they had such names as Thrinaxodontidae,

---

Six "molar" teeth of an early cynodont. These teeth mashed food and reflect a high energy level in their possessors, enabling the cynodont line to overshadow and replace the less well equipped bauriamorphs.

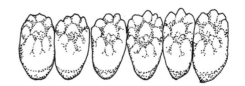

Diademodontidae, and Gomphodontosuchidae ("Trident-teeth," "Crown-teeth," and "Bolt-toothed Crocodiles," respectively)—appellations that suggest their owners' approach to more sophisticated chewing styles and consequently more nearly mammalian ways. This was a tremendous adaptive radiation, and each taxonomist has an individual way of divvying up the many results. All of these early Triassic cynodont families employed their advantages in usurping all econiches open to therapsid carnivores. They pressed the Bauriamorpha to extinction, finished off any remaining gorgonopsians and procynosuchids, and then radiated into a number of brand-new econiches, notably relating to insectivory, seed-eating, and other activities appropriate to lively small vertebrates.

The "Trident-teeth" are named for their representative genus *Thrinaxodon*, whose "molar" teeth are, indeed, tipped with three cusps, suggesting a well-developed shearing capability in these little (cat-sized and smaller) predators. In some classifications *Thrinaxodon* is included in a family Galesauridae ("Shark-lizards"), aptly named for a typical genus *Galesaurus*. Whatever family name is preferred by any given taxonomist, all members of this group were efficient little animals, very similar to mammals, whose lives were oriented around chasing insects and small lizards.

Thrinaxodonts display innovation in the structure of their "molar" teeth.

Skull of the insectivorous cynodont *Thrinaxodon*, showing shearing teeth and foramina about the snout region, which some paleontologists suspect may indicate the presence of whiskers and thus a nocturnal way of life.

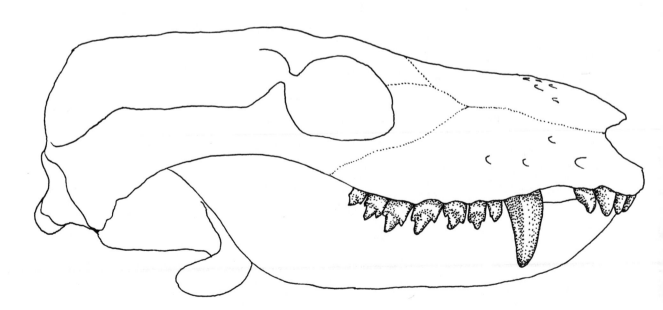

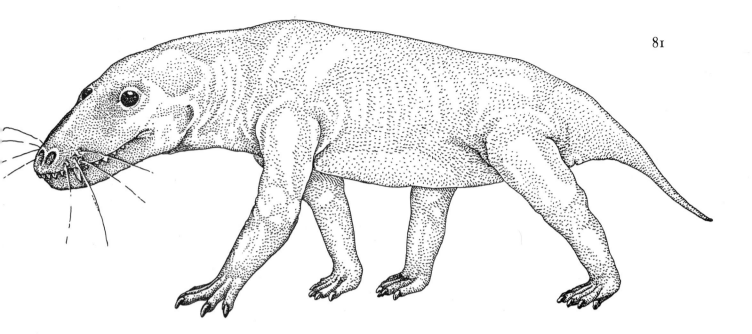

Restoration of *Thrinaxodon* as a nocturnal insect-eating predator with vibrissae.

These triple-tipped blades sheared past one another, upper over lower, to reduce food to small bites for quick, efficient swallowing. In more primitive therapsids these back teeth were constantly replaced as they wore out. Tooth replacement occurred in a wavelike sequence that lasted the life of the animal but often left gaps in the "molar" tooth row. For an animal with pointed teeth, this was no problem; but in the case of an animal with shearing molars, a missing tooth in the middle of the row was similar in effect to a big nick in a scissors blade: cutting could not take place at the gap. While such a problem is merely annoying for someone trying to use defective scissors, it could be fatal for a cynodont dependent on the efficient chewing of small bites of food. In connection with the selective pressure resulting from this problem, we see in thrinaxodonts and more advanced therapsids a reduction in the number of tooth replacements, so that disturbance of the molar row was gradually min-imized. This development represents a step toward elimination of the old system of perpetual tooth-replacement (or polyphyodonty, as some are pleased to call it) in favor of a single tooth-replacement (diphyodonty), result-ing in a set of permanent teeth, as in mammals. In cynodonts, it permitted improvements in chewing never before experienced by vertebrate animals.

In addition to their secondary palates and shearing molar teeth, thrinaxo-donts show further advances in food-processing suggested by the presence on the cheek areas of their skull of foramina (little holes), through which nerves and blood vessels passed. Such equipment very likely supplied nourishment to active lip and cheek muscles, which in turn may have supported vibrissae (whiskers), sensory structures enabling them to feel their way about at night. Because they were so small, thrinaxodonts may have found it safer to move about under cover of darkness, where their growing energy efficiency might serve to heat them as they searched out the insects on which they fed.

Now, the possible presence of vibrissae on these little cynodonts is highly significant, for the embryonic origins of whiskers are similar to those of hairs in general. In primitive ectothermic amniotes such as the living reptiles, the skin conducts heat; these animals are dependent on outside temperatures in maintaining their internal temperatures, and must be able to soak up or dissipate heat as fast as possible. Thus their skin is dry and thin, conserving moisture in the body by means of waterproof scales while controlling heat transfer through blood vessels close to the scaly surface. We may justifiably assume (since no trace of skin covering remains in fossils) that the more primitive end of the therapsid line possessed a similar integument serving a similar purpose.

No one knows exactly how hair was derived from such skin surfaces. Some

---

The ribs of *Thrinaxodon* overlapped, giving the animal some protection against the bites of larger predators. This specialization sets the animal to the side of the evolutionary line leading to primitive mammals, whose ribs do not overlap.

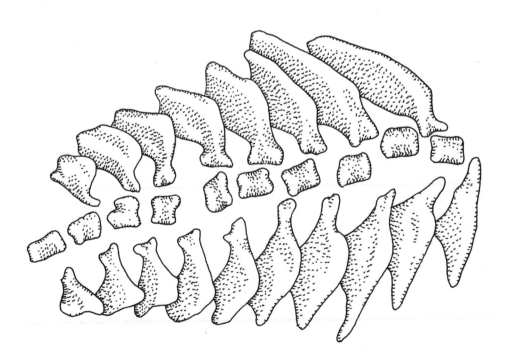

Probelesodon ("Before-dart-tooth"), a small, advanced carnivorous cynodont lying very close to the therapsid-mammal line.

students have suggested that it evolved directly from the keratinous (horn) scales of primitive therapsids, much as did the feathers of birds from the scales of primitive archosaurs. However, unlike either scales or feathers, which develop from the epidermis, hair arises from a deeper layer of the skin. The presence of those little holes on the cheek areas of some cynodonts suggests that thick nasal skin may have supported projections similar to those on some living snakes, projections that were perhaps antecedent to the honest-to-goodness whiskers of the sort we find on cats and mice. The gradual spread of such whiskerlike appendages across the cynodont integument not only augmented their efficiency at detecting insect prey in the dark but also served to trap air near the body surface and thus insulate the little animals against the nighttime cold.

While we will probably never know exactly when or how hair appeared, we may logically guess that it evolved early in the history of the cynodonts in the form of whiskers, and that some bauriamorphs possessed similar whiskers. The eventual displacement of bauriamorphs by cynodonts and the subsequent cynodont adaptive radiation may well have been due to the spread of these whiskers across their bodies to warm them; perhaps the bauriamorphs never made this crucial step toward complete endothermy.

Teeth and restoration of the "bolt-toothed" cynodont *Massetognathus* ("Chew-jaw"). This was an agile little animal whose generic name reflects its ability to grind up its food with its sharp-edged, irregular molars.

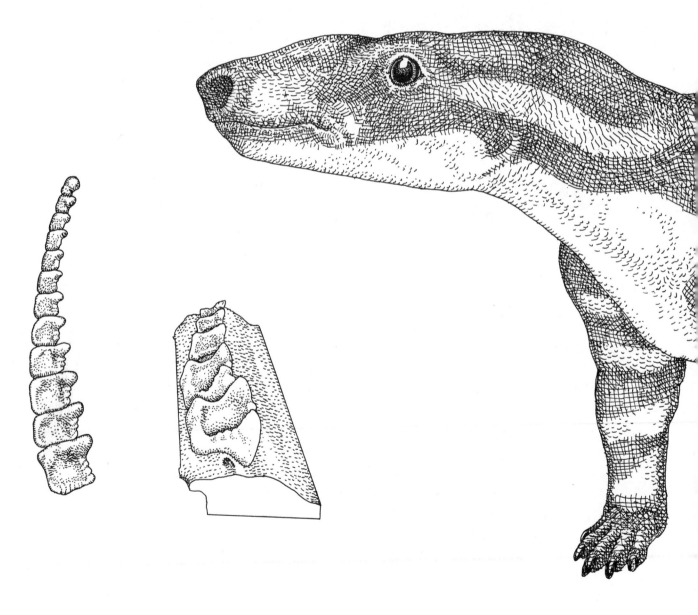

Their advantages permitted cynodonts an early Triassic adaptive radiation into new econiches based on small-scale vegetarianism. These herbivorous cynodonts are loosely lumped together under the name gomphodonts ("bolt-toothed ones"), and all had complex molar teeth useful in the grinding up of high-energy vegetable matter such as primitive seeds. Encountering no competition in their new econiches, gomphodonts became the most abundant of cynodonts during the middle of the Triassic period, around 210 million years ago.

The plentiful rat- to rabbit-sized gomphodonts fueled in turn lots of bigger cynodonts, perhaps the best known of which were members of family Cynognathidae ("Dog-jaws"), among which were some predators weighing

up to 90 kilograms (about 200 pounds). This carnivorous family is named for the type genus *Cynognathus*, whose dentary (tooth-bearing lower jaw-bone) angles sharply below the level of the rest of the jaw. The angle served as an area of insertion for a large, powerful masseter (chewing muscle) originating at the cheekbone. Cynognathids were ancestral to, and replaced by, members of family Chiniquodontidae, cynodont carnivores of rat- to dog-size that dominated the therapsid fauna of the Upper Triassic.

If we are to consider progress in the direction of mammals as advancement, the chiniquodontids were the most advanced, apparently, of the cynodonts; one genus, *Probainognathus* ("Progressive-jaw"), appears to straddle the traditional boundary line between therapsids and mammals. In such species as this we are again troubled by the vagaries of taxonomy in dealing with fossil material. As we mentioned before, paleontologists customarily divide mammals and therapsids at the point where the "reptilian" quadrate-articular jaw joint is replaced by one located at the juncture of the squamosal (rear cheek-

Three *Cynognathus* finish up a dicynodont dinner. Such "dog-jawed" predators grew to the size of bears.

bone) and the dentary. In *Probainognathus* we find an advanced form with both kinds of joints side by side, a real monkey wrench for the strange workings of taxonomy. To further complicate matters, the chiniquodontids are not the only therapsids to have sat astride that boundary.

A therapsid infraorder sometimes called Ictidosauria shares with the chiniquodontids a position so close to mammals that it appears to straddle the boundary between the groups. Most ictidosaurs appear to have been herbivores. Their teeth suggest a life-style very similar to that of small (rat-sized) modern rodents. In fact, the family Tritylodontidae ("Three-knob-teeth"), often included among the ictidosaurs, so successfully occupied the ecologic niche of small gnawing creatures that it survived far beyond the rest of the therapsids in time, living well into the Jurassic period up to about 100 million years ago. Some researchers have suggested that the living monotremes, the platypuses and echidnas of Australasia, are derived from tritylodont ancestors, although this is hard to prove because monotremes are so very highly specialized. Monotremes themselves (their name means "single-holed": their intestinal, genital, and urinary canals open into a single cloacal outlet like that of living reptiles) sit just barely this side of the reptile-mammal boundary,

and they are probably derived from a therapsid group different from that which was to produce the rest of the mammals.

We have traced the therapsids from their origins among pelycosaurs of the early Permian to the threshold of the origin of mammals in the Upper Triassic. But the therapsid dominion was based on the old rain-fertilized plants, and during the Triassic the rise of a totally new world of plant life was pushing these more primitive plants nearly to extinction. Cone-bearing and flowering plants were replacing seed ferns and their allies as the energy base on which land life subsisted.

Therapsids produced the first of a series of balanced land faunas which were to inhabit the earth. However, as "experimental models" (as all living things are) they left a few things to be desired designwise, especially when faced with the Triassic shift in the makeup of the world's flora. And as is always the case in evolutionary tales of this sort, someone else was waiting in the wings to take over.

Oligokyphus ("Few-humped"), a small tritylodont, feeding on a cone. Because of their chewing specialization, tritylodonts outlived all other therapsids.

THE MESOZOIC

During the period of greatest therapsid progress, along the Permo-Triassic boundary, there were many other nontherapsid animals experimenting simultaneously with the landlubbing life. Among these was a group of conservative swamp-dwelling reptilian characters rather reminiscent of (and ancestral to) modern crocodiles in life-style and appearance. Because their teeth were set deep in sockets in their jaws, we have named these animals thecodonts ("socket-teeth"), and give them their own order in a "reptilian" subclass Archosauria ("Rulers of Reptiles"). Archosaurs originated during the early Triassic as fish-eaters, but drying trends in the world climate forced some of them out of the water.

This is not to say that all thecodonts were forced out. Many of them were able to remain in the swamps to become the dominant inhabitants thereof for millions of years thereafter. Among their descendants are numbered the crocodiles, alligators, and gavials of today, which have persisted virtually unchanged in their swampy econiches for some 200 million years. In addition, a similar group, misnamed phytosaurs ("plant-lizards"), messed around in

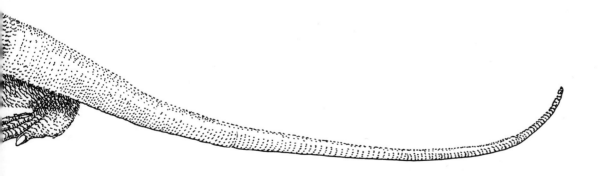

Chanaresuchus, a predatory thecodont of the Triassic of South America. *Chanaresuchus* and its thecodont kin were a vanguard of the archosaurs, which would ultimately squeeze the therapsids out of existence—and in so doing, squeeze from them the mammals as one might squeeze toothpaste from a tube.

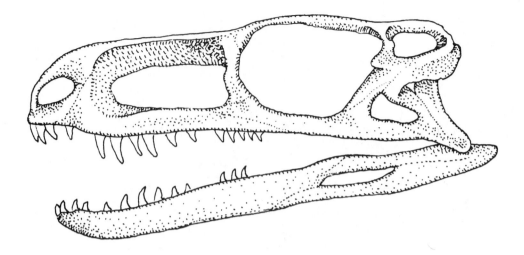

The diapsid skull of a thecodont. This is a light structure, a toothed trap and little else. Archosaurs were highly energetic animals that got around the problems of chewing by evolving gizzards in which to mush up their food.

the mud for many millions of years before being replaced by their more efficient crocodilian cousins. However, the really spectacular end of thecodont evolution took place on land, as the little insect-eaters improved on their locomotion in the search for fast-moving insect food. Again, insects fueled a revolution.

Early thecodonts looked a lot like modern crocodiles. Anyone seeing a crocodile out of the water will notice that it has unusually powerful hind legs and a long, flexible tail with which it sculls itself along in the water. On land, the crocodile appears somewhat clumsy, although it may move quite quickly if need be. The early thecodonts that were forced out of the water by the dwindling of their swamps appear to have been the smaller models whose movement on land was not too much affected by great weight; these gradually underwent a change in diet from fish-eating to insect-eating in response to their change in habitat. To do so, these early thecodonts capitalized on their powerful hind legs for bursts of speed to snatch insects or to escape predators such as big therapsids, and were gradually able to broaden their horizons.

Thecodonts had another advantage. They had a good, clean bite. You will perhaps recall the "apses" of the amniote skull we discussed back there somewhere, by which certain early landlubbers were able to improve their gapes and bites. Well, ancestral aquatic thecodonts had adopted a diapsid ("two-arched") condition and a correspondingly long set of jaws with which they might more easily nab the quick, slippery fishes that were their prey. These swiftly snapping jaws became even more valuable to the little thecodonts that took to chasing insects on land. So successful were they, in fact, that the group began to force its way into a wide variety of ecologic niches.

The thecodont spread was initially limited to the warmer parts of Pangaea, but advances in gait and temperature control later permitted these animals to reach cool climates and make contact with the established therapsid faunas of the southern end of the great primordial continent. In so doing, thecodonts expanded their appetites to include a lot of animals besides insects—animals of gradually increasing size. Animals such as therapsids.

Among the first thecodonts to tend toward large size were such predators as *Erythrosuchus* ("Red-crocodile," from the color of its fossil bones), a three-meter-long trotting hunter of heavy build whose way of life appears to have involved to a considerable extent the eating of our ancestors the therapsids. *Erythrosuchus* was so constructed that its gait was fully erect, rather like that of a modern mammal, a great improvement on the "three-quarter-pushup" stance of the therapsids on which it preyed. And so, this animal could outrun therapsids; and not only that: it could eat faster as well. Its skull was composed mainly of powerful jaws designed to shear great hunks of flesh with powerful bladelike teeth rather like those of some primitive therapsids.

But here the resemblance ended. Instead of chewing their food as did advanced therapsids, *Erythrosuchus* and other archosaurs simply bolted it down. They needed no molars, having evolved their parallel in the form of a gizzard, a "food mill" in which meat was tumbled about with stones to mush it up for digestion. Such an arrangement permits quick release of food energy for active animals, and allowed thecodonts an adaptive radiation based on sheer speed.

Equipped with gizzards, thecodonts appear to have entered quickly into competition with predatory cynodont therapsids during the Lower Triassic. They also ate these cynodonts, reducing their kind to small chiniquodontids and other comparatively marginal groups, much as the European invasion of the North American continent reduced the indigenous cultures to a very few specialized groups.

The real thecodont revolution took place in a suborder Pseudosuchia ("Fake-crocodiles") of animals whose hind legs became more and more important in their hunt for fast insects. They started out as small, light forms whose posture suggests a growing ability to rear onto their hind legs for quick dashes after their prey, which they knocked down with their forepaws. Many such animals, included in the aptly named family Ornithosuchidae ("Bird-crocodiles"), became more or less completely bipedal as their need for speed

increased. This change freed their forelegs from the tyranny of being walked on, and the forelegs in consequence became organs of manipulation— "hands"—with which prey might be seized and torn apart.

By the time of the Middle Triassic, such animals as *Ornithosuchus*, from which the family receives its name, and *Gracilisuchus* ("Slender-crocodile") were becoming the dominant predators of Pangaea. Running erect on their hind legs, they so far outran, outate, and otherwise outmoded the remaining therapsids that the rest of the Triassic became a single long, sad decline for the latter group.

Thecodonts did not stop with the elimination of the larger therapsids. Placing pressure on one another and evolving alongside the rise of the new plants and the new insects of the Mesozoic, these advanced animals quickly diversified to fill the old therapsid econiches. By the middle of the Triassic such innovative little archosaurian predators as coelurosaurs, the earliest dinosaurs, had evolved from thecodont ancestors to prey on insects and the remaining little therapsids. Coelurosaurs, whose name means "hollow-lizards," were swift, hollow-boned, birdlike beings, lightly built bipeds that were originally about the size of house cats or chickens. They were supremely agile runners with large, acute eyes, and long, stiff tails balancing their trunks at the hip, forever freeing their forelimbs to make havoc with insects, small therapsids, and other, lesser animals.

From coelurosaurs descended the rest of the mighty dinosaurs, an astoundingly diversified and successful group of archosaurs, whose dominion of the earth's continents spanned most of the Mesozoic era's 160 million years and

---

A Triassic confrontation between a therapsid and a bipedal thecodont called *Gracilisuchus*, illustrating the differences between therapsids and thecodonts that would permit the replacement of the former by the latter. The advantages of bipedalism (speed and raptorial forefeet) allowed thecodonts to spawn the dinosaurs in a great adaptive radiation that lasted through the rest of the Mesozoic era.

whose descendants still rule the skies in the form of birds. Dinosaurs filled almost every terrestrial econiche in which we now find mammals. Not only did they produce a wide variety of light predators in the coelurosaurian mode; dinosaurs were represented by an incredible array of big flesh-eaters ecologically similar to the great cats and wild dogs of today, all of whose life-styles were based on the art of getting about swiftly and efficiently on two legs.

A primitive dinosaur, the New Mexican coelurosaur *Coelophysis* ("Hollow-bone"). In a world dominated by such efficient predators, the mammalian descendants of therapsids had no choice but to stay out of sight.

These terrible animals fed on a wonderful diversity of plant-eating dinosaurs, which in turn consumed anything from pine trees to poplar leaves. Some of these vegetarians grew to vast size, dwarfing the long-gone dicynodont therapsids and, indeed, all land animals before or since.

Like therapsids, dinosaurs have long been considered reptiles, and for many of the same reasons. However, a glance at a few dinosaurs suggests that they were quite *un*reptilian in the Linnaean sense, having been for the most

The Mesozoic era was a time of absolute rule by the dinosaurs. The unique mammalian consciousness enabling us to read this book originated during the Mesozoic in response to the dinosaurian dominion.

part far more erect and more active than therapsids or, for that matter, many mammals. And these specifically nonreptilian aspects of dinosaur biology, retained in their living representatives, the birds, spelled the end for the therapsids and relegated their mammalian descendants to a marginal ecologic position throughout the rest of the Mesozoic.

In retrospect, it may logically be said that mammals were made out of therapsids by dinosaurs. As we shall see, many of the basic mammalian characteristics evolved specifically as counters to the archosaurian triumph; we are in many ways reflections of the superb dinosaurian way of life through its unrelenting pressure on our long-gone ancestors. Although we may perhaps breathe a sigh of relief that the dinosaurs are (excepting birds) gone, we might pause to reflect gratefully on their part in molding our kind.

THE MAKING OF THE MAMMALS

The Triassic triumph of the dinosaurs was so complete that for the next 120 million years during the Jurassic and Cretaceous periods, descendants of the once mighty therapsid order remained an insignificant lot of little mammals all smaller than house cats. Because of the delicacy of their bones and their choice of a forest environment unsuited to the process of fossilization, few traces of these meek ancestors of ours survive today. These remnants consist mostly of fossilized teeth—the hardest parts of the body—and the jaws in which they grew. However, we are able to learn a lot about the habits and appearance of our Mesozoic progenitors because of the importance of mammalian teeth in processing food. We could even regard teeth as "graphs" of the evolution and "shape" of mammalian econiches; such thinking tends to make one want to brush them after every meal and at bedtime. The great evolutionist George Gaylord Simpson dubbed mammals "glorified reptiles." Having waded thus far through this book, we might put it this way: mammals are dinosaured therapsids. What, then, were the changes forced on those therapsids—poor, stupid things—by the archosaurian ecologic victory? How were these momentous changes accomplished?

We left the therapsids in their Triassic twilight, during which pressure from the rising dinosaur dynasty had reduced them in variety to a few insect-eaters and specialized vegetarians. All of these, Triassic forebears of Caesar himself, were rather ratty in appearance. Like rats, they were lively, energetic beings whose living was made at night. Caesar, incidentally, is said to have remarked on natural selection: *Sic pilum iactum est* (That's the way the javelin is thrown)—but it is unlikely that he was referring to therapsids at the time.

Being small animals, the last therapsids lived in a spatially labyrinthine environment where even a pebble could be an obstacle. They were forest-dwellers, and their world was one of tree trunks, roots, fallen branches, holes, dinosaurs, and other interruptions to peace of mind. Because they moved under cover of darkness, they couldn't see much of their surroundings and were under ever-increasing pressure to enhance their senses other than sight. You can appreciate the problems faced by these little animals by imagining yourself living in the following "econiche":

You are blindfolded such that you can discern light and dark, but no more. You live in a vast junkyard piled three and four meters deep with old automobiles, interspersed here and there with vast pits and the ruins of tall buildings. Not only is this Hell inhabited by packs of man-eating wild dogs, but in the air helicopters circle, filled with armed men who hope to shoot you. Your appetite is tremendous; to survive, you must eat your own weight in food *each day*. The only food is live rabbits, which you must catch with your own hands. Not only must you escape predators and snare food in this nightmarish landscape; you must also find a mate, raise a family, and leave descendants. Our Mesozoic ancestors were heroes, every one.

The "blind-in-the-junkyard" econiche certainly offers selective pressure aplenty, doesn't it? You and your descendants would have need for an almost telepathic awareness of your environment to survive. Precisely this happened during the Mesozoic, when our forefathers and -mothers evolved tremendous sensitivity to sound, touch, and odor to make up for their near-loss of sight as a result of nocturnalism. Such sensory changes necessitated corresponding alterations in the brain—in consciousness itself. The fundament of human intellect lies deep in the Mesozoic era.

As we have seen, the original vertebrate brain was a three-part expansion of the front end of the nerve cord. Each of the three bulges accommodated sensory input for one important sensory modality: smell for the forebrain (prosencephalon), vision for the midbrain (mesencephalon), and balance and detection of vibration for the hindbrain (rhombencephalon). In fishes and amphibians, the functions of these bulges became more elaborate as their possessors occupied more varied econiches. Nonetheless, the ancient three-part brain persists throughout the vertebrate world as a "neural chassis" on which higher developments of the brain are superimposed like layers in an onion.

The structure of the brain in therapsids approximated that of reptiles living today. The original three-part brain lay deep in a lumpy structure reflecting millions of intervening years of active landlubbing. Sophisticated behavioral and physiological temperature control, bodily coordination, and sensory sys-

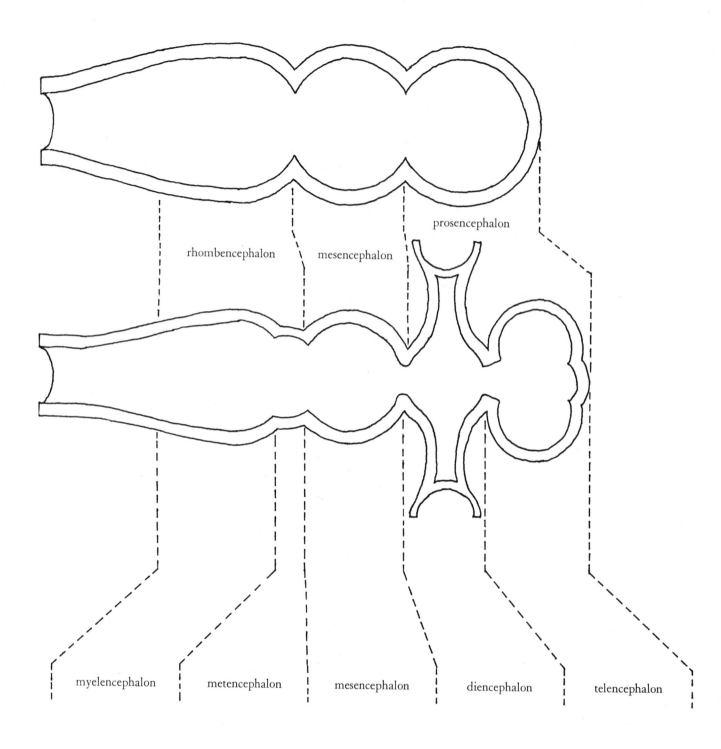

rhombencephalon

mesencephalon

prosencephalon

myelencephalon

metencephalon

mesencephalon

diencephalon

telencephalon

Early in vertebrate evolution, the old tripartite brain subdivided into five different functional areas that persist today. From front to rear, these parts function, roughly speaking, as smell and association center (telencephalon), visual integration center (diencephalon), primary visual center (mesencephalon), balance and coordination center (metencephalon), and center for control of viscera (myelencephalon).

tems were linked together in this brain by inborn behavioral "maps" of the way therapsid life was supposed to be. Like modern reptiles, then, therapsids were probably rather automatic beasts whose acts were dictated by the experiences of thousands of ancestral generations. They were gifted at birth with the ability to care for themselves, hunt appropriate prey or eat appropriate plants, recognize and escape predators, and otherwise function in a world they had not yet seen but for which they were nonetheless prepared genetically.

In reptiles we are able to see this sort of life-style in the flesh. Such common reptiles as snapping turtles never see the hatching of their offspring. The young turtles are about as large as fifty-cent pieces and are relatively soft of shell and essentially defenseless. Nevertheless, these infants make their way unassisted to water and immediately set about catching insects, small fish, and other animals, or seeking out carrion. Of any twenty baby snappers, perhaps two survive a year or more to adopt the peculiarly ferocious ways of their parents.

That this mode of living is successful is evidenced by the fact that the family Chelydridae, to which snappers belong, has persisted almost unchanged for some eighty million years—but at the cost of any sort of progressive mental development. Because of their choice of environment, snapping turtles do not experience significant competition for energy or space, a situation that easily lends itself to the truculent Republican durability of snapper ways. Our own ancestors, on the other hand, had to contend with dinosaurs. These dramatically successful archosaurs *ate* therapsids and mammals—with a sanguinary enthusiasm shown by precious few predators toward snapping turtles.

Archosaurs were originally daylight predators on insects (once they gave up fish), and their brain cavities suggest great development of visual centers in the midbrain, coupled with intricate coordination of balance and high rates of activity in a complex hindbrain. Menaced by predators such as these, the rigid inborn behavioral patterns of therapsids were forced to give way to change so fundamental as to require a totally new sort of brain. This new brain specifically reflected in its adaptations the pressure of archosaurs. It is our brain, in which the ancient memory of dinosaurs can still evoke a chill of fear.

Getting back to your junkyard econiche, you have to find food while blindfolded. Therefore you must spend much of your time being quiet and listening for the scurrying of those rabbits. Because rabbits are fast, you have to locate them precisely and pounce on them, using only your ears to pinpoint the little devils. Our ancestors experienced an identical need for fixing on sounds heard

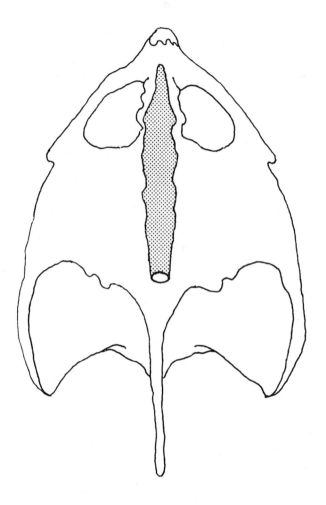

In living reptiles the brainpower is limited and the life-style consequently rather specialized. The skull of a snapping turtle with the brain volume shaded.

in the space "out there." This was a space not visually perceived; it had to be "reconstructed" *inside the brain*. This necessitated changes in the wiring of the hindbrain.

The hindbrain originated as a swelling in the nerve cord that coordinated bodily response for the system of orientation with which the first vertebrates stayed upright in their weightless watery medium. Connected with this sense of balance and orientation was the lateral-line system, an apparatus for perceiving vibration—sound—in the water. So important were these systems to our first vertebrate ancestors that their hindbrain centers gradually enlarged, eventually producing a distinct cerebellum ("little brain"), which, while having grown progressively more complex as vertebrate life evolved, remains even today mainly a center for integration of sensory input and bodily orien-

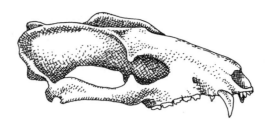

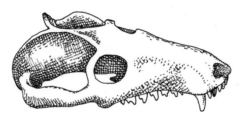

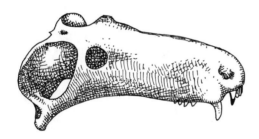

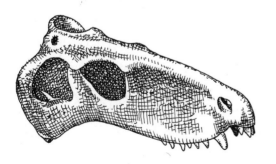

The growing strength and dexterity of the bite in the synapsid line is reflected in the growth of the temporal opening in which the jaw muscles operate. *From bottom*: skulls of a pelycosaur, a primitive therapsid, an advanced therapsid, and a mammal.

tation. Beneath the cerebellum lay the hearing centers, originally alerting the cerebellum directly to respond to sounds perceived from the environment.

The hearing equipment of the synapsid line went through some interesting evolutionary changes before producing its highest exemplar in the ear of Johann Sebastian Bach. The nature of these changes was first suspected by the German anatomist C. Reichert, who in 1837 suggested that the tiny bones in the inner ears of people and other mammals were developmentally identical with certain bones in the jaws of reptiles. Specifically, Reichert showed that two of these minuscule bones, the malleus (hammer) and incus (anvil), arise during embryonic development from the same structures from which the bones of the jaw joint (articulate and quadrate) evolve in reptiles; we could say that these bones are homologues of one another, that they share their identity in our common ancestry with reptiles. The story of the gradual change of occupation of these bones, from jaw articulations to hearing accessories, is really a lot of fun.

Early tetrapods such as pelycosaurs possessed mandibles (lower jaws) composed of a dentary (tooth-bearing bone) associated with a number of thin plates of bone that originally served to strengthen the jaw as a whole. The hindmost of these bones, the articular, articulated (hence its name), or hinged, with the quadrate bone of the skull. In therapsids this arrangement persisted, but the new life-style of these progressive animals, with its growing emphasis on high consumption of energy for lots of activity, required precise manipulation of food in the mouth in order that it might be better broken down for efficient digestion. In these animals, the act of chewing was forcing changes in the structure of the lower jaws to strengthen them further and increase their biting leverage.

In this process, the dentary bone was becoming longer and thicker in response to the increase in stress to which it was subjected. In addition, as the molar teeth came into being and were put to grinding up food, changes in jaw musculature forced changes in jaw conformation to permit good chewing. The most primitive therapsids had jaws whose muscles were mainly attached to the inner surfaces of both jaws and skull. In later models, however, new masseter (chewing) muscles gradually appeared on the outside of the dentary bone, connecting this to the cheekbone. Other muscles arose from a new coronoid (crownlike) process that extended the rear of the dentary bone upward inside the arch of the cheekbone. These muscles inserted along the top and back of the skull, along the temporal bones, and hence are collectively called temporal muscles. Because of the increasing power of these muscles in

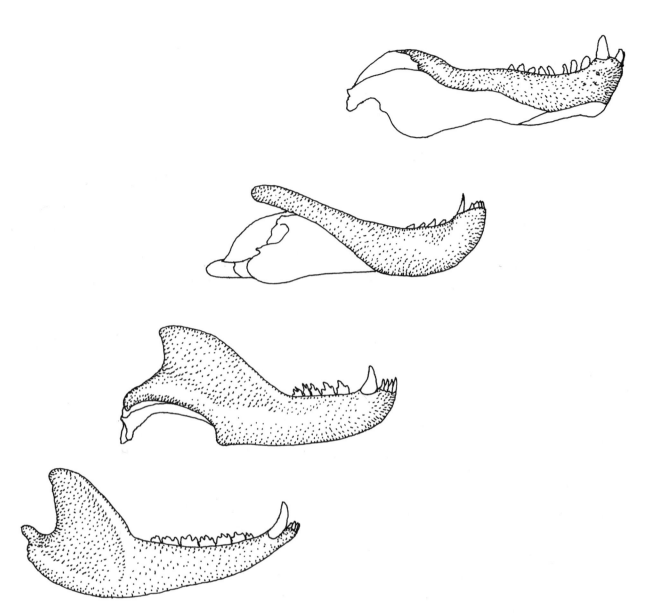

As biting strength and chewing ability increased through synapsid history, the dentary (tooth-bearing bone) of the lower jaw responded by growing—eventually to become the only bone of the lower jaw. The dentary is shaded in this view (*from top*) of the lower jaw of a pelycosaur, a primitive therapsid, an advanced therapsid, and a mammal.

---

higher therapsids, the dentary bone was becoming bigger and heavier all the time at the expense of all the other little jawbones.

Now, we recall that in the long-lost days of the Pennsylvanian and early Permian, our ancestral line was represented by the pelycosaurs, a low-slung crew indeed. Because of their posture, these animals spent a lot of time with their heads, and thus their lower jaws, resting on the ground. Which had its

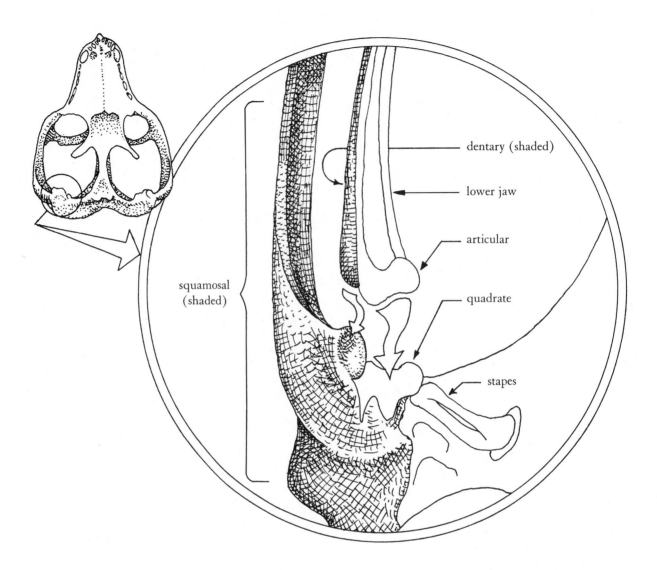

The evolution of the mammalian jaw joint and ear from the therapsid condition. *Left*: an advanced therapsid jaw, with both the old articular-quadrate jaw joint and a new dentary-squamosal joint. *Right*: a mammalian jaw, in which there is a dentary-squamosal joint only, the dentary being the only bone in the lower jaw; the articular and quadrate have migrated into the middle ear and retain their old sound-transmitting functions.

advantages. The ears of these early tetrapods appear to have been as yet inefficient at picking up airborne vibrations such as those we call sound. For these early landlubbers, fresh from the water, the only necessity for hearing was related to detection of the footfalls of potential predators, "sounds" that are transmitted well enough by the bones of the lower jaw from the ground to the balance organs of what we now call the inner ear.

These balance organs were (and still are) fluid-filled structures equipped with nerve ends highly sensitive to movement of the fluid. In early tetrapods,

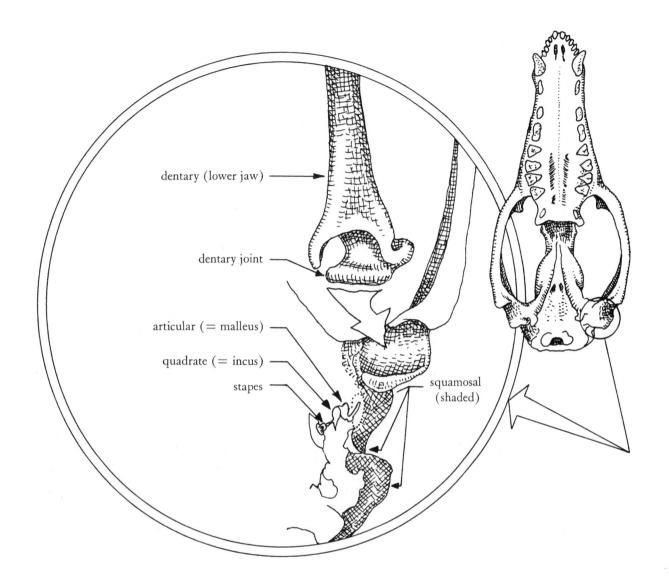

dentary (lower jaw)

dentary joint

articular (= malleus)

quadrate (= incus)

stapes

squamosal
(shaded)

extensions of these balance organs came to be associated, in the form of the hyomandibular bone, with part of the old tongue-support mechanism in fishes. Originally supporting part of the upper-jaw rims with the tongue, this little bone aided the transmission of vibration to the balance organs. As land vertebrates got better at detecting airborne vibration through their jawbones, the hyomandibular was reduced in size to become the stapes (stirrup), connecting the tympanum (eardrum) to the cochlea (snail-shell-shaped inner ear). Air vibrations moving the tympanum thus moved the stapes, which wiggled the fluid in the cochlea, which alerted the brain.

This is the way the ancestors of reptiles, archosaurs, and birds took to hearing, and it served them very well. The therapsids, however, were forced along a different route as their hearing improved. As the subtlety of land-lubbing increased, synapsids continued for a time to keep their jaws to the ground in hopes of picking up noise; in them the tympanum stretched along the rear of the jaw in such a way as to be intimately associated with the articular bone. As the dentary bone increased in size, however, a conflict arose: increase in chewing efficiency forced an increase in dentary size at the expense of the little articular bone, which was still necessary for hearing. Increased chewing efficiency threatened hearing acuity: how would it work out?

Ultimately, the little jaw-joint bones, the articular and quadrate, receded into the head as cynodont evolution progressed. With the inception of the mammalian jaw joint—that is, one between the dentary mandible and the squamosal cheekbone—these tiny bones were entirely freed from their function with relation to the jaw. They continued to be jointed together, however, and their ancient articulation with one another persists as they transmit sound from the eardrum to the cochlea. Here they remain, and here we hear, still listening with parts of our mouths in the manner of our ancestors 300 million years ago.

This arrangement in the highest therapsids and first mammals permitted good, sharp hearing. But it was inherently no better than the stapes system of the archosaurs until some method was evolved for the hearing system to triangulate accurately on noises heard in the dark. Thus evolved the uniquely mammalian pinnae (fins), or external ears. The gorgeous auditory appendages of jackrabbits, deer, cats, dogs, and most other mammals probably stem in all their variety from simple flaps of skin arising at the rear of the jaws in therapsids and serving to protect the tympanum from punctures. With the move into night-hunting, these became sound-collectors that concentrated, funnellike, air vibrations on the tympana. Like radar screens, these collectors were adjustable in direction, and information about the position of sound-sources relayed to the brain permitted their precise location in the dark. The external ears of most mammals continue in their function of "focusing" sound into the tympanum, and thus are "lenses" analogous to those of the eyes. Our progenitors "saw" in the dark, with their ears!

Meanwhile, back in the midbrain, vision was being set more and more aside. As a result, areas of the brain devoted to vision experienced a sort of "evolutionary neglect" during the early evolution of mammals. Being very small animals, mammals had correspondingly small eyes. Small eyes contain little room for the large numbers of retinal cells necessary for acute vision. Being lovers of darkness, the early mammals had no need for good color vision at all, and lost most of the retinal cone cells that distinguish light wavelengths and permit most nonmammalian vertebrates to see color. Even today, most mammals do not perceive color well, even those to whom, as in

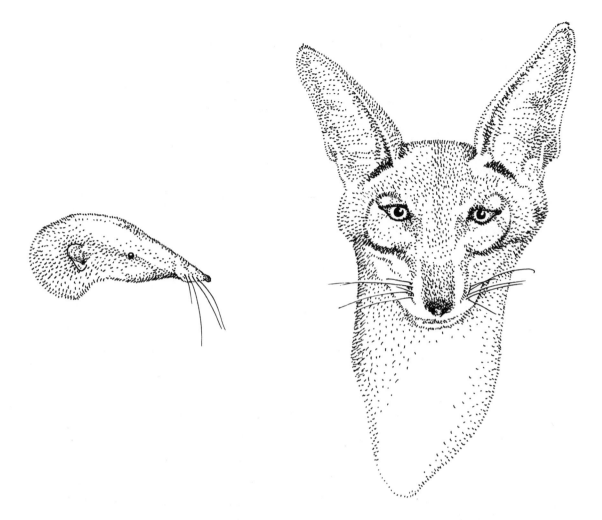

From the sound-locating pinnae of the primordial mammals (*left*) descend such lovely structures as the external ears of the living black-backed jackal (*right*), which can pinpoint small prey in thick cover with their aid.

cats, the eyes have experienced a resurgence in importance. Conspicuous exceptions include ourselves among the primates—but more about our color vision later.

The forebrain of the last therapsids was subjected to the most notable changes of all as these small animals receded into the night. Originating as a pair of olfactory bulbs in early fishes, the forebrain had differentiated into twin structures of two major parts each: an olfactory lobe, the location of odor sensation; and a cerebral hemisphere, in which bodily response to those odors was initiated. As vertebrates took to landlubbing and more complex lives, the cerebral hemispheres became important centers for the correlation of all sensory input with bodily responses. As their importance increased, these hemispheres experienced a corresponding increase in volume. In such animals as amphibians and primitive reptiles the cerebral hemispheres comprise three sections: at the bottom of each lies a volume occupied by the basal nuclei, a region through which sensory information and instructions for muscles are relayed to and from the hindbrain; these basal nuclei are covered by two pallia ("cloaks"), together making up what we call the cerebral cortex.

Oldest of the "cloaks" is the paleopallium (literally, "old cloak"). It is a center for distinguishing smells, and to a certain extent retains this function in living mammals. Above the "old cloak" lies the archipallium ("primary cloak")—at least in primitive tetrapods of today. This was probably also the condition in the more primitive therapsids. The archipallium is a center of correlation of impulses from the sense of smell with those of other senses, and as the vertebrate line progresses, it gets bigger to handle the exigencies of progressively more complex life-styles. Like the paleopallium, however, the archipallium remains most intimately connected with the sense of smell, integrating this important equipment with the rest of the being in which it is situated.

As we have seen, as the therapsids evolved and exerted ever more pressure on one another, they radiated into new niches dependent on increasingly complex behavior and sensory awareness. To handle the greater needs for correlation of increasing input to the brain, these animals evolved greater volume in their cerebral hemispheres. At some point in their evolution there appeared a wedge of tissue between the fore-ends of the paleopallium and the archipallium. This was the neopallium ("new cloak"), which in its original form was useful in correlating data gained through the sense of smell, by means of which appropriate response during hunting was initiated. Later, the neopallium evolved into the most spectacular of mammalian specializations. The origin of the neopallium probably coincided more or less with a change in nose structure experienced by small, advanced therapsids that were learning to hunt at night. In more primitive animals, the openings for the nostrils occur on the sides of the skull, well apart from one another. As some of them took to messing around in the night, however, these nostrils gradually moved closer to the tip of the nose in order to expose more of the odor-sensitive lining of the snout to the environment directly in front of these snuffling animals.

The area of the lining of the nose simultaneously experienced an enlargement among higher therapsids, thus permitting them many more individual olfactory (smell) cells than had been available before. This was accomplished by the evolution within the nose of paper-thin turbinal bones, scroll-shaped structures whose convoluted surfaces exposed huge numbers of "smell-cells"

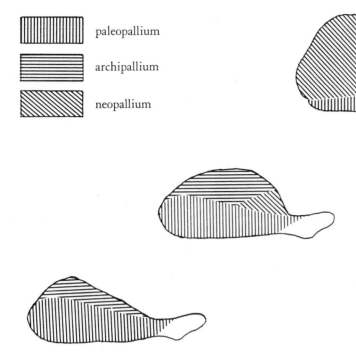

paleopallium

archipallium

neopallium

olfactory bulb

The origins and growth of the neopallium. *Left*: a primitive therapsid cerebral hemisphere; *middle*: that of an advanced therapsid or primitive mammal; *right*: that of a more advanced mammal.

---

to the air, in turn permitting clearer odor messages for processing by the evolving neopallium.

The sense of smell is a very strange thing. Although there appears to be no limit to the different odors that our noses can distinguish one from another, the cells that sense these odors are all fundamentally identical. This is in strong contrast to the condition in our other chemical receptor, the sense of taste, in which four different kinds of cells distinguish four tastes (sweet, sour, bitter, salt) only. Somehow, those identical smell-cells convey to us all the subtle gradations of odor in the world.

The distinguishing of these myriad smells must take place in the brain rather than in the nose itself, and as the sense of smell became more and more important to Triassic therapsids, there occurred corresponding changes in their braincases to permit better odor discrimination. This dependence on the sense of smell for one's living presents an interesting set of problems.

Daytime predatory therapsids hunted their prey much as plant-eaters found their food: they simply spotted that prey in the distance, trotted up to it, and

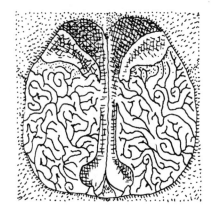

End-on view of the nasal opening in a mammal, showing the edges of the scroll-like turbinal bones, which serve to increase smelling acuity.

ate it. Their "eating sequence" was "turned on" by the mere sight of a prey animal, and they did little real hunting beyond a haphazard wandering around. Their nocturnal relatives, on the other hand, were unable to see *anything* at any distance beyond a few centimeters. When the nocturnal hunters located an odor signifying the presence of their favorite prey, or bigger animals that might nab them instead, they were still separated from the source of the odor by gaps in both *time* (that is, the interval between the moment when the scent was released, at a distance, and the moment the hunter perceived it) and *space*. Such gaps had to be filled in if the premammals were to contact and eat prey, avoid predators, even find a mate. Taking to hunting with these nonvisual senses about which we've been reading, they were forced to seek *traces of* their prey in the form of molecules of odor instead of seeking the prey *itself*. Those night-hunting ancestors of ours, in short, had to perceive their world at second hand!

Inside their heads evolved analogues of their frighteningly unpredictable world in the form of an "inner space" in which odors and sounds replaced vision. Here, within their little heads, distances in time and space were bridged by *planning*. Sequences of discrete behaviors were strung together like beads to bridge the gap between odor-of-prey and the prey itself, and these sequences varied to suit the vagaries of the invisible world without. So appeared the first purposeful use of a sense of time and a new awareness of consequences.

In this inner space, one molecule might signify a tasty cockroach, another molecule might warn of the silent presence of an archosaurian predator. Each such molecule was a matter of life and death, and each came to produce a powerful "image" of appropriate "form" in the inner space of our nocturnal ancestors. Although we human beings are primarily visually oriented mammals, most of us have experienced a tremendous rush of vivid imagery associated with certain smells that had some importance in our pasts. A smell can recall to us faces, rooms, even powerful emotions, in a reawakening of the ancient Mesozoic thought-patterns from which our self-awareness springs.

Now, these "images" are certainly not any result of direct visual perception of the source of smells. No, indeed! They are instead *reconstructions* of the source—understanding at one remove. The making of a molecule into a meal or a monster, and then bridging the time-space distance to that meal or monster in appropriate response, required all sorts of cerebral additions. These took place in the neopallium, the organ of the inner space, during the long dark of the origin of mammals. Symbolic thought was born in the terrible Mesozoic night, and what was to become the human spirit arose in the nose.

The smell-illuminated inner space was so important to our Mesozoic forebears that its organ gradually increased in volume to handle its increased capabilities. At first merely a little fold of tissue at the front of the cerebral hemisphere, the neopallium bloated out across the entire brain, permitting ever finer tuning of the inner awareness to the outer reality. Simultaneously, hearing became the central means of locating movement, direction, and volume within the inner space, so that sounds, too, became translated into imagery. Even now, we often recall intense emotions, scenes from the past, and other poignant sensations simply by hearing a sound such as a piece of music with which we once associated those sensations—testimony to the intimacy with which hindbrain auditory centers are connected with the inner space within the neopallium.

Inner space saved us from the archosaurs, and the subsequent history of mammals is characterized by a general capitalization on the use of this unique faculty. In higher mammals the neopallium is so large that its surface is deeply convoluted to contain the vast number of neurons of which self-awareness is composed. We human beings are so highly specialized in the use of this Mesozoic holdover that it sometimes gets in our way, tripping us up by concocting such awful things as nuclear weapons and bureaucracies. But in this inner space lies all that is art, too, and music, and gourmet cooking, and books about therapsids. And deep, deep down in that inner space, which now reaches out to the edges of the universe, deep in some dreadful corner, the archosaurs still stride about like great mad birds from whose glittering, stony eyes we must at all costs remain hidden.

Because the inner space was an indirect way of dealing with the world, it came to take more and more time for the Mesozoic night-dwellers to get their minds "in gear" with the outer world. Associating molecules with events depended in large part on experience, a period of trial and error during which the student sniffer was subject to all sorts of perils. And so, selective pressure gradually mounted for some sort of parental protection in higher therapsids to set their young off on the right track in an uncertain world. Luckily, by the time of significant neopallium growth there was probably already a measure of parental care of eggs resulting from various pressures that had acted on egg-laying therapsids from the very beginning.

The built-in automatic behavioral system of primitive vertebrates tends to occur hand in hand with a distinctive mode of reproduction in which the main "investment" is in the number of young, rather than in the quality and quantity of care devoted to each individual. This is a very efficient mode of reproduction as long as its practitioners are not subject to too many environmental changes—recall our friends the snapping turtles, specialized aquatic reptiles with heavy armor that inhabit an ecologic niche that has persisted almost unchanged for millions of years. Because their young are not too bright, and must look out for themselves, snappers lay lots of eggs.

Similarly, early therapsids probably invested lavishly in number of offspring but almost not at all in any relationship with eggs or young. However, the very nature of eggs resulted in pressure among these therapsids for a bit of attention to their well-being. Eggs are highly fragile objects, the most vulnerable interval in an egg-layer's life cycle. An egg cannot flee danger, nor can it strike out in its own defense. Also, eggs are nutritious and delicious; there are many animals, including the author of this book, who eat eggs whenever they

Brains of a therapsid and a primitive mammal. In the therapsid (*above*), the brain is essentially tubular and the parietal eye may be seen poking out at the top. In the mammal (*below*), the neopallium has produced a wrinkle over the olfactory lobes and the brain fills most of the rear of the skull.

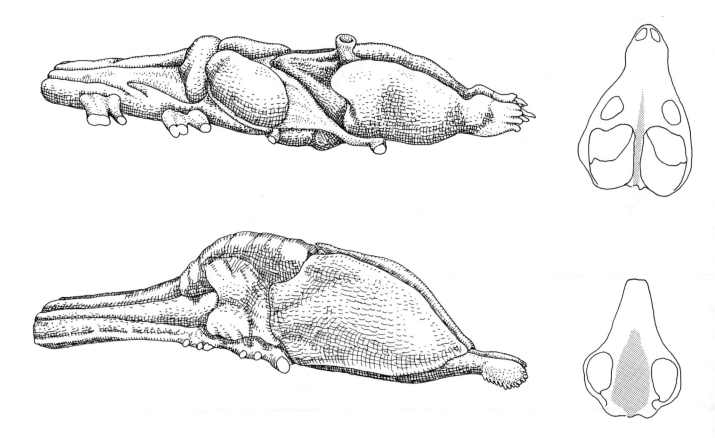

can. This was as true back in the Triassic as it is today, and there gradually evolved among therapsid mothers a tendency to remain with their eggs in a burrow until they hatched out. We can see this behavior today in the monotremes, egg-laying furbearers who are sometimes described as the only surviving therapsids.

From this level of maternal concern it was but a short step to care of the young after they had hatched. Precisely when this occurred we will never know, but it may have coincided with the evolution of higher operating temperatures among therapsids. Because of their small size, newborns among warm endothermic animals, whether birds or mammals, are far more subject to death by loss of heat than are their parents. In such situations happiness is a warm mother, and therapsid females that stuck with their eggs brought warmer, healthier young into the world. In defending their young against predators, the pioneers of maternal behavior further enhanced their progeny's chances of survival. Real *motherhood* made an entrance in biological history.

Another selective pressure bearing on reproduction resulted from the growth in complexity and efficiency of the teeth in advanced therapsids. You'll no doubt recall that the more primitive therapsids shared with modern reptiles a mode of tooth replacement in which, throughout the animals' life spans, new teeth erupted along the jaws in waves, popping up between older teeth and replacing these as they wore out. Thus many of these animals were rather snaggletoothed most of the time, with mouths containing full-sized teeth alternating with small new ones or with toothless spaces. This arrangement is fine if all you do is grab your food and swallow it, but what if you need to break it down first, to mush it up for faster energy release, as the higher therapsids found it necessary to do?

We have already traced the process in which these animals experienced gradual reduction in the number of tooth replacements through their lives in order that gaps in the tooth rows might be eliminated. This change helped, but it created new problems for the very young. Because they were chewers, they needed precisely opposed sets of teeth from birth. Also, a chewing life-style necessitates relatively powerful jaws, in response to which, as we have seen, jaw mechanics and leverage improved as the cynodonts evolved. But great jaw power was incompatible with the small size of the head in newly hatched therapsids, as was the requirement that these little hatchlings have complex, adult-style teeth. So, selective pressure arose again to force therapsids— probably the small, insectivorous models in particular—to augment parental egg-tending with some sort of parental feeding of the young. In response to this pressure there appeared the essence of mammalian life in the form of the feeding of toothless, dependent young on secretions from the skin of the adult female.

The skin of higher therapsids was probably hairy and moist already, having responded long before to the selective pressure for temperature control. In these animals the secretions maintaining the suppleness of the skin and permitting fine control of body temperature were probably similar to those of modern mammals. Such secretions contain a variety of fats, salts, and other nutrient-rich materials that, during some momentous interval in the career of our therapsid ancestors, became an important food source for the young.

In present-day egg-laying monotremes, an early form of this skin-feeding is retained (as are so many primitive traits in these animals) in the form of aggregations of tiny sweat-gland-like ducts whose openings are interspersed with long, stiff hairs. It is these hairs that the young monotremes lick (they do not suck on nipples as higher mammals do) for their first nourishment. The liquid produced by the skin glands is rather thick and yellow, its thickness permitting it to cling to the hairs until it is licked off by the young. It is a fatty compound containing proteins but lacking true milk sugars of any sort. This "milk" is probably similar in composition to that of the first milk-producing therapsids, in which oily fats that lubricated the mother's skin might have served the newborn as food. Those innovators of mother love still used all their blood sugar for themselves, rather than sharing it with their young in the form of milk sugars as higher mammals do.

The advent of suckling further tightened the bond between mother and offspring, eventually permitting the young to test their growing inner space against the real world while still enjoying the protection of an experienced parent. This constellation of adaptations relating to life in the dark permitted the last Triassic therapsids to circumvent the advances made by dinosaurs in

---

The highly specialized living monotremes of Australia and New Guinea offer us a look back in time to the transition between the last therapsids and the first mammals. Egg-layers with primitive milk glands and therapsidlike limb structures, they may be more closely related to the extinct multituberculates than to other mammals. If so, their line of evolution is likely to be quite separate from that of the rest of mammalian creation, suggesting that such "mammalian" attributes as fur, efficient temperature control, and the feeding of young with milk may have been present in therapsids themselves. *Above*: a platypus (*Ornithorhynchus*, "Bird-nose") of Australia, a powerful swimmer that feeds on small aquatic animals and hides her eggs at the end of a ten-meter tunnel in a riverbank. *Below*: an Australian echidna, or spiny anteater (*Tachyglossus*, "Quick-tongue"), a powerful digger that feeds on termites and ants, and carries her eggs in a temporary abdominal pouch.

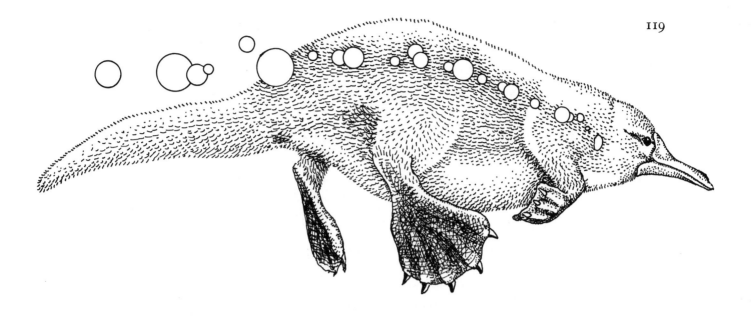

the fine art of living. But in so doing, these therapsids ceased being therapsids and entered a long darkness during which they were, more than anything else, rough beasts slouching toward mammalhood. For them the rest of the Mesozoic would be a time of refinement, of perpetual trial.

By the end of the Triassic, about 185 million years ago, at least three orders of "nominal mammals"—those whose jaws articulated through a dentary-squamosal joint—are known to have existed. (They may also be regarded as specialized therapsids, depending on how we feel about them at any given time.) Each of these orders has been granted a thunderous multisyllabic name longer than any of the little animals belonging to it; because these dawn mammals were so small and fragile, little besides their teeth remains to intrigue the modern observer, and to these teeth they owe their magnificent names: Multituberculata ("Many-lumped-teeth"), Triconodonta ("Triple-cone-teeth"), and Symmetrodonta ("Symmetrical-teeth").

All of these impressive names describe small animals that to a casual observer would be almost indistinguishable from modern shrews or mice. The largest Triassic mammals may have reached the comparatively great size of ten centimeters (about four inches) in length, but their teeth all suggest diets based on the rapid chewing and swallowing of such high-energy delicacies as insects and seeds. In addition, we find among these tiny creatures evidence, in the form of diphyodonty, with deciduous ("milk") teeth, that a period of true infancy was already well established among them and that suckling was standard procedure. Some taxonomists have therefore suggested that the presence of diphyodonty be regarded as equal in importance to the development of the jaw joint as a criterion for establishing a dividing line between therapsids and mammals.

Like living shrews, these first mammals were little more than moving snouts. Their heads were long and narrow, and equipped with rows of complex teeth serving the triple functions of seizing, shearing, and grinding. These highly efficient tribosphenic (triple-wedged) teeth were necessary in such diminutive animals, whose energy expenditure simply to keep warm was tremendous. Living shrews may consume several times their body mass in a day's eating, and although the earliest mammals may have been somewhat more moderate than this in their diet requirements, we may guess that they ate a great deal. Primitive modern mammals such as tenrecs and hedgehogs, insectivorous beasts related to shrews, are nocturnal animals that conserve a bit on food requirements by maintaining body temperatures a bit lower than the norm for mammals. Nonetheless, they, too, have ungodly big appetites.

In the Mesozoic world of vast evergreen forests and bright-eyed archosaurs,

The primitive mammal *Morganucodon* ("Morning-tooth") of the Upper Triassic, whose entire appearance reflects the exigencies of its nocturnal, smell-oriented life-style.

---

the early mammals lived invisible lives, guided by their long snouts and night-brains through a complex, three-dimensional world of roots and fallen trees. In addition, they experienced relentless pressure to be quiet. The fossils of the small, meat-eating dinosaurs of the sort that ate mammals indicate very acute hearing on the part of these birdlike predators, and it behooved our mammalian forebears not to betray themselves to those archosaurs by making noise.

In order to move swiftly but silently across the forest floor, the primordial mammals abandoned the somewhat splaylegged walk of their therapsid ancestors in favor of a bounding gait in which the body flexed in a vertical plane, up and down, rather than sideways. At the same time, the feet of these first mammals moved beneath their bodies, rather than to the sides, making them better able to flit into the narrow spaces that were their salvation from predators.

All this required alterations in the structure of the spinal column, in which there appeared a set of specialized lumbar (loin) vertebrae along the lower back. In therapsids, ribs spread all the way from the shoulder to the hips, a pair for each vertebra; in mammals, the lumbar vertebrae have no ribs and

ribs along lower back, no lumbar vertebrae

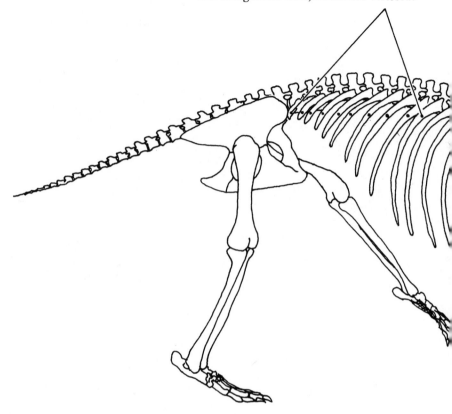

The skeletons of an advanced therapsid (*above*) and a mammal (*below*), showing important differences between the two. In every way, the mammalian skeleton suggests greater flexibility and activity.

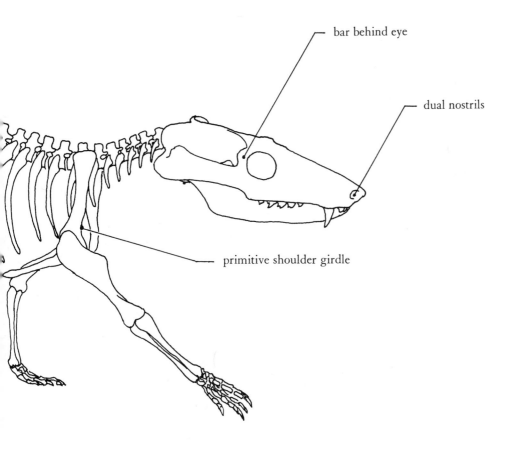

bar behind eye

dual nostrils

primitive shoulder girdle

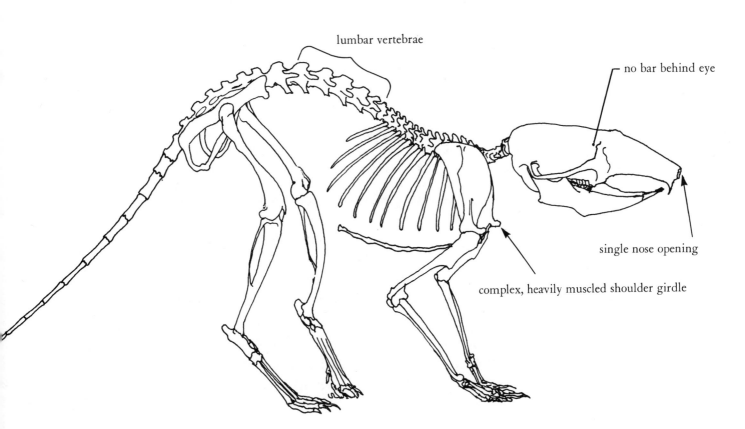

lumbar vertebrae

no bar behind eye

single nose opening

complex, heavily muscled shoulder girdle

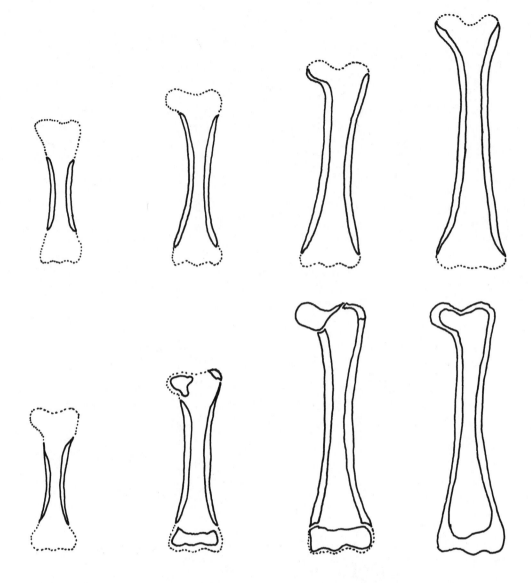

Diagram of different bone-growth patterns in therapsids and mammals. *Above*: the therapsid pattern, in which bone growth (solid lines) begins at center of shaft, gradually replacing cartilage (dotted lines). *Below*: the mammalian pattern, in which bone growth occurs at the middle and at both ends of the bone shaft; when these different areas of growth meet (*third from left*), the epiphyses fuse with the shaft and growth is complete (*at right*).

are free to flex up and down. This change allowed our ancestors to move in the bounding, hopping gait characteristic of mouse-sized mammals today.

Early mammals experienced another change in response to the exigencies of a miniaturized life. In most therapsids and other more "reptilian" animals, the life cycle was one of continued growth until death. Certainly, this growth slowed in adulthood and nearly ceased in old age, but it was not the precise cessation to which we are accustomed in mammals. Mesozoic mammals, on the other hand, experienced such unrelenting pressure to remain small that they evolved a skeletal growth process sharply ending at adulthood, ensuring an adult life of constant small size.

This change was aided in part by a change in the mode of growth of the long bones of the limbs. In embryonic vertebrates, these bones are first formed of rubbery cartilage; as life proceeds, the cartilage is gradually replaced by bone in a process that begins at the center of the bone shaft and proceeds gradually toward the joints, which remain composed of cartilage. In mammals, the long bones grow not only from the center toward the ends, but from the ends toward the center as well. The end-growth centers, epiphyses ("bone-ends"), permit the young mammal use of efficient, strong joints of bone rather than cartilage, even as the bones continue growing; when the growth from the bone center meets that of the epiphyses, all elements fuse and growth abruptly ceases.

Meeting no competition in the dark world they inhabited, and well fed by the diversity of insects and seeds therein, primeval mammals bore many young and diversified. The Multituberculata, having evolved a set of gnawing teeth comparable to those of modern rodents, gradually replaced the remaining therapsid tritylodonts, whose ancestors had independently evolved similar characteristics. Perfecting their gnawing way of life, multituberculates became the longest-lived order of mammals, surviving almost 130 million years until true rodents independently evolved a better gnawing apparatus long after the fall of the dinosaurs.

The multituberculates and their shrewlike symmetrodont and triconodont cousins were each independently derived from insectivorous therapsids during the Triassic. In the next geologic period, the Jurassic of 185 million to 130 million years ago, these three orders were joined by two more: Docodonta ("Beam-teeth") and Pantotheria ("All-beasts"). Docodonts joined symmetrodonts and triconodonts in an early Cretaceous extinction, but the pantotheres, named for their position at or near the trunk of the family tree of living higher mammals, were destined to do great things.

Like most of their mammalian contemporaries, the pantotheres were shrewlike animals. At their level, taxonomists have founded the subclass Theria ("Beasts") of class Mammalia. Theria are divided in turn into three infraclasses: Pantotheria, Metatheria, and Eutheria. In this arrangement we see the founding by the pantotheres ("all-beasts") of two higher groups: metatheres ("between-beasts," the living marsupials) and eutheres ("true-beasts," the placental mammals, including ourselves).

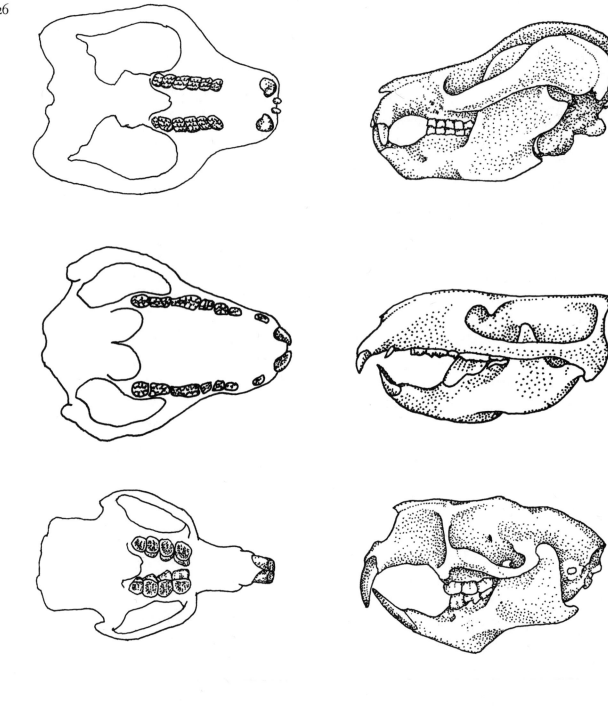

Three separate but parallel tooth specializations for gnawing vegetation. In each case, the specialization permitted its possessor to achieve great success. *Top*: a tritylodont therapsid skull, viewed from underneath and from the side. *Middle*: a "mammalian" multituberculate of the sort that replaced tritylodonts during the Jurassic. *Bottom*: a modern rodent, the beaver. Rodents replaced the multituberculates during the early Cenozoic and are the most numerous of mammals today.

Both metatheres and eutheres bear live young, a fact suggesting that this practice originated somewhere in their common pantothere ancestry. Live birth was yet another step in the therapsid-mammal path of increasing intimacy between mothers and offspring, a path that began long before with egg protection and was enhanced along the way with the evolution of nursing from milk glands. As the selective value of skin secretions for the feeding of young became apparent, improvements were made in their passage to said young. Nipples came into being, a great leap forward in that the milk no longer simply seeped onto the coat of the mother but was retained inside her body until released by a sucking action on her nipples. In this way, less milk was wasted and more efficient energy transport between mother and young was initiated.

A further step in the process of increasing intimacy between mother and young occurred when the eggs were retained within the mother's body for a time, during which she was freed from the necessity of protecting them and was able to increase her collection of energy in preparation for their hatching. A similar process occurs in many living reptiles such as the common garter snakes of genus *Thamnophis*, whose young are hatched within the mother and are born alive, able to care for themselves at birth. Garter snakes are thus spared adverse selective pressure on a set of fragile eggs, which the defenseless mother snake could have no hope of protecting.

Refinement of this practice of carrying the eggs internally came, probably among the pantotheres, with the elimination within the mother's body of the eggshell, which was no longer necessary to protect the embryo. This permitted direct contact between mother and young before birth, and in time both metatheres and eutheres took advantage of that contact by eliminating the yolk sac, formerly necessary to nourish the developing embryo, and replacing it with some form of direct exchange of nutrients and wastes between mother and embryo. This exchange takes place through an organ called the placenta, an incredibly complex network of blood vessels joining mother to young. The embryo, in other words, lives as though it were an organ of the mother's body until it is born. So momentous was this form of development that the descendants of its meek pantotherian originators quickly took over the mammalian world.

The first metatherians of which we are aware appeared about eighty million years ago in the form of animals nearly identical to their modern descendant, the North American opossum. Indeed, the humble possum *Didelphis* ("double-uterus") itself originated in those distant dinosaur-dominated times, and in its conservatism may be said to be one of the most successful of mammals. But the opossum's success, and that of all metatheres, is limited by the manner in which it bears its young.

Among opossums, up to twenty young are born at an undeveloped stage, whereupon they crawl to their mother's teats and attach themselves, remaining thus fixed for seventy days. At the end of this time, their development is approximately equal to that of newborn eutheres, or placental mammals, in whom nourishment of the unborn young is far more complete. This discrepancy between metatherian and eutherian reproduction permitted the eutheres an early lead in the Cretaceous evolution of mammals.

Were it not for the fact that the continents of Australia and South America broke away from the rest of the world's landmass during the Cretaceous, eutheres might have driven their metatherian relatives into extinction during the Mesozoic. As it was, the two island continents were peopled originally by metatheres of the single order Marsupialia ("Pouched Ones"), which radiated adaptively across them to produce a complete fauna of metatherian mammals parallel to the more advanced eutherian fauna of the northern continents. Although much of that fauna disappeared from South America when it rejoined the northern world and was invaded by advanced placentals during the Cenozoic era, Australia remained pretty much separate from the placental world until the advent of human beings and their many satellite animals. Australia's metatherian fauna has independently produced a whole marsupial world that survives in the form of a delightful diversity of marsupials from kangaroos—the grazing equivalent of the deer and antelope of the rest of the world—to Tasmanian devils, a group of marsupial "wolverines."

The early placental eutheres were so successful, within the limits imposed by dinosaurian hegemony, that they quickly diversified into a number of groups exploiting heretofore unoccupied ecologic niches in the Cretaceous world. Most primitive of these were the conservative order Insectivora, the insect-eaters in whom the original specializations for night-living continue almost unchanged to this day. That these little animals are superbly secretive is suggested by the fact that, although they occur by the thousands in almost any rural landscape, they are almost never seen by their human relatives, the sharpest-eyed of mammals. The dinosaurs trained these little dynamos well.

In addition to superb senses of smell and hearing, insectivores have picked up on a number of interesting specializations as they improved their little acts. Some of them have poison glands, modified from salivary glands, with which they stun their prey. These glands occur in the North American short-tailed shrew *Blarina brevicauda*, and while harmless to human beings, they are more than adequate for killing prey, such as mice, bigger than the ever-hungry shrew. In addition, many shrews are equipped with a nasty stench, making them unattractive to mammalian predators. This accounts for the large numbers of shrews killed but left uneaten by family cats.

While some of the insectivorous hunters stuck close to the ground, giving rise to the hedgehogs, moles, and shrews of today, others early took to the trees and, uncontested in that environment, diversified. There were no dinosaurs in the trees, and the pioneer arboreal mammals found it advantageous to come out in the daytime. Among living primitive tree-dwellers may be mentioned the tree shrew *Tupaia*, a squirrellike creature whose ancestors may have been among the first mammals to abandon the scaly therapsid tail in favor of a glorious furry rudder with which the agile little climbers might guide their leaps from branch to branch.

Emerging into the daylight at the forest crown, the new arboreal mammals evolved a sense of vision that remains today without parallel among mammals. In their world of leafy green, they needed a sense of color to spot the insects and fruits that comprised their diet. They also needed a good sense of depth with which to gauge their leaps through an aerial three-dimensional world, and to this end stereoscopic vision appeared among them, coupled with grasping paws in which one digit opposed the rest for precise seizing of branches. Such a congeries of adaptation made for unmatched tree-dwelling capability and resulted in the foundation of our own order Primates ("First") among those squirrellike insectivores of the Cretaceous jungles. They weren't much to look at, but among their remote descendants would be the only earthly creatures able to read this book.

Along with the rodentlike multituberculates, the insectivores, and the marsupials with which they shared their world, these primal primates were pretty much the only expressions of mammalhood for the last twenty or so million years of the Cretaceous period. Excepting the primates, these animals were, for the most part, still living the lives of the nocturnal therapsids from which they were descended, probably still sharing among them that scaly tail whose function remained (and is today) related to heat dissipation and temperature control.

The late Cretaceous, biologically speaking, was the best and liveliest of times. The flowering plants, having long before evolved a symbiosis with insects in which the plants exchanged food for transportation of reproductive material, had come into their own by the Middle Cretaceous, and the ensuing millions of years saw the world looking much as it does now, at least with regard to plant life. In response to this triumph of flowers there had evolved an animal world of sumptuous diversity and, to modern eyes, startling outlandishness.

Across the sunlit plains roamed vast herds, bisonlike, of horned dinosaurs whose teeth could chop up wood itself. Representing a mighty plant-harvesting ecologic force, these big grazers were complemented in the forests by flocks of browsing bipedal archosaurs, some of which were sheep-sized, others of which weighed as much as elephants. In conifer forests, sauropod dinosaurs grazed on the tall trees, their long necks swaying gracefully, giraffe-like, in the dappled forest light. All these magnificent vegetarian dinosaurs were kept in shape by a fantastic crew of archosaurian predators that walked erect on two legs, ever watchful for a sick adult or untended young one among the herds of plant-eaters. Some of these meat-eaters were huge, big enough to tackle the heavily armed horned dinosaurs, while others were merely man-sized, swift runners that chased the speedy lesser herbivores among the dinosaurs and sliced them to pieces with knifelike talons.

There were smaller meat-eating archosaurs, too, which fed on insects and on whatever mammals they could catch. But the mammals, having perfected their synapsid inheritance, were doing very well in spite of their attentions. Although none of these Cretaceous mammals was much bigger than a cat, they had all been honed by 120 million years of dinosaur oppression into a fine-tuned crew of inner-space sniffers. Represented by the rodentlike multi-tuberculates, marsupials, forest-bottom insectivores, and tree-dwelling primates, these furry creatures shared with a growing number of birds the art of being small and living in microhabitats in which larger animals could have no interest.

In the oceans, a similar diversity characterized the Cretaceous living world. The floating microscopic plants called plankton fueled fishes and sea turtles similar to those with which we are familiar today. Feeding on these was a host of wonderful whalelike and seallike animals related distantly to modern lizards. In short, the Upper Cretaceous was a time of unparalleled diversity. It was a time of consolidation, in which new econiches were exploited by plants, their insect allies, and all the rest of life dependent on them. It was a time of celebration.

Then something vast and terrible happened. What it was, we will probably never know; it was so sudden that it left no record of its nature, except in one way: all of the biggest, most conspicuous, and most energy-intensive animals in the world disappeared from the fossil record at the end of the Cretaceous, from the oceans as well as from the land. So complete was this death that with

it we end the Mesozoic era and begin a new one, the Cenozoic, the Era of the
New Animals.

Whatever ended the Mesozoic may have struck at the world's energy base,
photosynthesis, in which sunlight is locked into energetic molecules by plants.
This guess is borne out by the fact that all the larger animals—the ones that
had to keep eating all the time to stay alive—were the first to go. Either they
sat at the tops of food chains as active carnivores, or they were plant-eaters
that consumed large quantities of food at all times to fuel their energetic life-
styles. A temporary dimming of the sunlight reaching the world's plants would
kill just such animals first. The plants, however, especially those that could
"hibernate" in the form of seeds, or otherwise store energy through bad times,
might gear right up again when the crisis passed. And so they appear to have
done.

For the world's animal life, though, it was a different story. In the oceans,
the reptilian "whales" and "seals" disappeared, along with certain large fishes
and other animals. On land, all archosaurs—with the significant exception of
birds, which inhabit microhabitats between which they can fly—ceased to
exist. But their relatives the semiaquatic crocodilians, slowgoing folk with a
good ability to fast and hibernate, survived, along with reptiles such as turtles,
lizards, and snakes, themselves accomplished fasters. Mammals, too, survived.
Their small size enabled them to hide out in their microhabitats, and many of
them were able to slip into the torpor of hibernation and fast away the hard
times as did the crocodilians and reptiles.

What could have dimmed the sunlight reaching our planet, if indeed this is
what happened? Could a nearby supernova have raised the earth's temperature, filling the sky with dense clouds for a year or two? Could a great me-
teorite have hit our world, darkening the face of the land for months with dust
from its annihilation? Whatever it was, could it happen again? Who knows?
Certainly not I.

One of the nice things about life as a whole is its ability to thrive on
misfortune. The monstrous catastrophe that ended the Mesozoic era sixty-five
million years ago left the world a lonely place, stripped of all its large land-
dwelling vertebrates. With the exception of a few aquatic econiches occupied
by conservative types such as turtles and crocodiles, the seas, too, were devoid
of large air-breathers.

There were, however, plenty of birds and mammals left to survey the
wreckage. These, all sporting endothermy and high activity levels, simultaneously made bids to occupy the thousands of vacated large-animal
econiches. The Paleocene ("Old Recent") epoch opened the Cenozoic era

with tremendous adaptive radiations by birds and mammals to fill out the depleted terrestrial ecosystem. For a time these groups competed with one another as carnivorous flightless birds two meters (about six feet) tall stalked the plains like caricatures of the dinosaurs before them. But these birds lacked the grasping forefeet of their departed cousins, and they had no teeth. The mammals quickly defeated them and filled up all the old dinosaur econiches in a process of diversification that continued through more than sixty million years until the advent of our own kind about three million years ago.

In addition, mammals filled the marine econiches vacated in the catastrophe. In the oceans, the mammalian inner space continued to function as an analogue of that largely invisible aquatic world, and in response the neopallium became a tremendous convoluted organ with which these seagoing forms "saw" their environment by reading echoes of sound pulses.

Mammals even tried for the air, with the neopallium serving a similar function in echo-reading during night flight. Bats appeared very early from insectivore ancestors, and while they are confined to the night by competition from the more adaptable and sharper-eyed birds, they are excellent fliers in their own right.

On land, the inner space permitted the evolution of all sorts of new behaviors, and, consequently, new econiches. Because of their teaching and learning capabilities, mammals were flexible in their behavior to a degree that was new to the world. Predatory animals took to stalking, to plotting out complex maneuvers by which they might nab their wily prey; the prey in turn evolved stratagems to evade such threats; and thus the brains of carnivores and herbivores paralleled one another in a growth spiral that continues to this day.

Meanwhile, in the trees, the old primate order remained in many respects a conservative, generalized group. However, in one respect they were progressive: their way of life superimposed on the old inner space a new world of perceptions based on the three-dimensional, full-color living in the treetops for which they had become specialized. They evolved complex societies using vocal signals across the forest spaces, and they defended members of their troops from attack.

Between five and ten million years ago, a change in forest conditions forced some African representatives of the order out of the trees and onto the open land. Descendants of these fallen primates relearned the art of getting about on the ground, this time in the presence of great cats, dogs, and other specialized carnivores. Like their relatives in the trees, the new, ground-dwelling primates were sharp-eyed, and they ate almost anything. These were the earliest hominids, members of the Family of Man, and to their eyes, hands, societies, and omnivorous exploration of their world, we owe our own existence.

At some point, the hominids learned to exploit the energy locked into gigantic herbivorous mammals such as elephants and rhinoceroses. They did so by capitalizing on the planning capabilities inherent in the inner space to evolve tools and tactics, which were disseminated among their groups by the complex vocal signals for which living in the trees had prepared their ancestors. They became bipedal social hunters and experienced a brief adaptive radiation; but they also systematically killed one another off, using tactics devised in the neopallium to do so. Thus they exerted tremendous selective pressure on one another's neopallium for more and more efficiency at both plotting and at the interchange of plots from one individual to another. Language and culture appeared as extensions of the inner space; to accommodate these new specializations, the hominid neopallium grew right over the entire brain and created *Homo sapiens*, the best inner-space dweller of them all.

Our kind, whose toolmaking enabled it not only to construct analogues of the outer world in the inner space but also to reverse this process and impose conditions of the inner space on the outer world, quickly radiated into a host of ecologic niches. In the last three million years human activity has killed off some 70 percent of the larger mammals and will probably eliminate the rest within a few decades. Most of the world's biomass, the weight of its living matter, is being mopped up by human beings right now in a process suggesting that there will soon be nothing alive on the earth but human beings and the algae they'll have to be eating. Humanity represents the most complete adaptive radiation a single species has ever made on this planet, and, indeed, shows a capability for radiating right out into the wastes of extraterrestrial space, a serious option in light of the pressures coming to bear on us in our present limited environment.

The long synapsid story ends here, for we've caught up with ourselves. But the everlasting cycle of pressure–change–adaptive radiation–pressure continues as it has for billions of years. Understanding this cycle is not only useful in examining past history; it is crucial if humankind is to halt the current berserk planetary scramble for ever more efficient weaponry with which to continue the age-old competition for resources.

Ten thousand years ago, human beings had killed off most of the large mammals, and had begun to experience selective pressure for a change in diet. The change came, in the form of the human-plant symbiosis called agriculture; agriculture in its turn permitted an adaptive radiation of human cultures, which is right now facing the next step in the cycle: selective pressure from overpopulation. And after pressure must come change.

Will the change at whose threshold we stand be one that permits us to continue the mammalian way? Will it be a humane change, purposively introduced through conscious planning for the benefit of our kind? Or will we ignore the coming change, allowing it to steal upon us in the night in some malignant incarnation for which we will answer with our enslavement or extinction? And whose is the next adaptive radiation?

GLOSSARY

Agnath "Jawless," any of a group of primitive fishlike vertebrates that founded the subphylum Vertebrata during the Ordovician period, about 460 million years ago.

Amniote Of or pertaining to the amniote egg—that containing a sac of liquid in which the embryo may develop; the hen's egg is the most familiar example. Amniote vertebrates—reptiles, birds, and mammals—are those possessing such eggs.

Amphibian Any of a class Amphibia of "two-lived" vertebrates whose eggs are not protected from drying; amphibians are hence restricted to areas near water and represent an evolutionary level between fish and amniote.

Anapsid See *Apse.*

Anomodont Any of a suborder Anomodontia ("Nameless-teeth") of mainly herbivorous therapsids.

Anteosaur Any of a family Anteosauridae ("Prior-lizards") of large primitive carnivorous therapsids.

Apse An arch; in this context one of the bony arches of the skull beneath which the jaw muscles function. Anapsid = "archless," denoting the primitive condition in amniote skulls; euryapsid = "broad-arched"; diapsid = "two-arched," the condition found in most living reptiles and in dinosaurs and birds; synapsid is a misnomer for "fused-arched."

Archipallium The "primary cloak," a layer of the cerebral hemispheres lying directly above the paleopallium (q.v.) and serving to correlate sensory input from the nose with bodily responses.

Archosaur Any of a group of amniotes distinguished by possession of a diapsid skull and by powerful hind legs and a comparatively active life-style. Crocodiles, dinosaurs, and birds are members of the archosaurian line.

Bauriamorph Any of a therapsid infraorder Bauriamorpha of small, active, advanced carnivorous animals.

Caseid	Any of a family Caseidae of large vegetarian pelycosaurs.
Chelonian	Any turtle or tortoise, member of the reptilian order Chelonia.
Chordate	Any member of the phylum Chordata, possessing at some time during its life cycle a notochord, or stiffening rod, and a dorsal nerve cord.
Cochlea	The snail-shaped cavity in the inner ear through which sound impulses are transmitted to the nerve endings responsible for the sense of hearing.
Coelurosaur	Any of a group of small predaceous dinosaurs of birdlike construction and great agility.
Cotylosaur	Any of an amniote order Cotylosauria ("Stem-lizards"), comprising the earliest and most primitive forms from which subsequent amniote groups evolved.
Cynodont	Any of an infraorder Cynodontia ("Dog-teeth") of advanced, largely carnivorous therapsids.
Cynognathid	Any of a family Cynognathidae ("Dog-jaws") of advanced cynodont therapsid predators.
Dentary	The tooth-bearing bone of the mandible, or lower jaw; in mammals the lower jaw consists entirely of a dentary bone, while in reptiles the dentary is but one of several mandibular bones.
Dentary-Squamosal	In mammals the joint through which the lower jaw articulates with the cheekbone; in reptiles the jaw articulates through an articular-quadrate joint.
Diapsid	See *Apse*.
Dicynodont	Any of an infraorder Dicynodontia ("Two-dog-teeth") of advanced anomodont therapsids.
Diencephalon	A visual integration center in the advanced vertebrate midbrain.
Dimetrodon	"Two-measure-teeth," a genus of sailed predatory pelycosaurs.
Dinocephalian	Any of an infraorder Dinocephalia ("Horrible-heads") of primitive anomodont therapsids.
Dinosaur	Any of two orders of advanced archosaurs dominating land vertebrate life during the Mesozoic era and probably ancestral to birds. Dinosaurs occupied most of the econiches in which large mammals are found today.
Diphyodont	The condition of having only two sets of teeth (deciduous and permanent), as in mammals.
Docodont	Any of an order Docodonta ("Beam-teeth") of primitive Mesozoic mammals.
Dromasaur	Any of an infraorder Dromasauria ("Running-lizards") of small, agile anomodont therapsids.
Ecology	The study of the distribution of matter, energy, space, and time by the living system or parts thereof.

Edaphosaur   Any of a suborder Edaphosauria ("Ship-lizards") of herbivorous, often sailed pelycosaurs.

Emydops   "Resembling-a-turtle," a genus of small dicynodont therapsids whose mouths still retained a few cheek teeth.

Endothiodon   "Inside-yellowtooth," a genus of dicynodont therapsids whose mouths still retained a set of grinding teeth in the roof of the palate.

Epiphysis   "End-of-the-bone," the growth center in mammalian joints that permits firm articulation early in life; in other amniotes the ends of the bone are composed largely of gradually hardening cartilage, and bone growth proceeds only from the center of the bone shaft.

Erythrosuchus   "Red-crocodile," a thecodont (primitive archosaur) of the sort that came into conflict with the therapsid dynasty during the Triassic period, ultimately causing the complete extinction of therapsids except for small nocturnal forms.

Euryapsid   See *Apse.*

Euthere   "True-beast," one of the subclass Eutheria of placental mammals.

Glossopteris   "Tongue-wing," a genus of seed fern common in cool parts of Pangaea during the Permian.

Gomphodont   "Bolt-toothed," a condition in which true chewing molars occur among small advanced cynodont therapsids close to the therapsid-mammal boundary.

Gracilisuchus   "Light-crocodile," a genus of thecodont archosaurs of the sort that eventually put the larger therapsids out of business.

Haptodon   "Fastened-teeth," a small pelycosaur of primitive and generalized form from which more advanced pelycosaurs of various sorts may have evolved. *Haptodon* was an insect-eater.

Hemichordate   Any of a phylum Hemichordata ("Half-chordates") of wormlike animals closely related to primitive chordates.

Herbivory   The eating of plants as a way of life; vegetarianism.

Homology   Evolutionary identity between structures in different but related organisms; the forefeet of frogs, the wings of birds and bats, the forefins of whales, and the arms of human beings are homologous structures, each of which is a variant of the primitive forelimb structure of the first tetrapods.

Jonkeria   A genus of large omnivorous and herbivorous titanosuchian therapsids, among the most primitive therapsids to experiment with vegetarianism.

Lycosuchus   "Wolf-crocodile," a predatory therocephalian therapsid with saberlike canine teeth, of the Upper Permian.

Lystrosaurus   "Shovel-lizard," a genus of advanced dicynodont therapsids of southern Pangaea and Antarctica.

Mammal	Any of the class Mammalia of endothermic amniotes that feed their young from mammary glands. They are the only living therapsid descendants.
Mandible	The lower jaw of vertebrates.
Marsupial	"Pouched," any of the Metatheria.
Massetognathus	"Chew-jaw," a genus of advanced gomphodont therapsids of small size, lying somewhere near the therapsid-mammal evolutionary boundary.
Mesencephalon	In the vertebrate midbrain, a primary visual center.
Metathere	Any of the subclass Metatheria ("Between-beasts") of mammals whose young are born at an immature stage and suckled in a pouch or otherwise attached to the mother's teats; "marsupial."
Metencephalon	A balance and coordination center in the vertebrate hindbrain.
Microhabitat	A very small or narrow ecologic "space" or a way of life that does not make great demands on the environment. Microhabitats are ordinarily occupied by small animals and because of their diversity support incredibly varied faunas such as insects, birds, and small mammals.
Molar	In mammals and advanced therapsids a grinding cheek tooth.
Monotreme	"Single-holed," any of the primitive mammals of Australia and New Guinea that still possess a reptilian cloaca through which intestinal, reproductive, and urinary tracts open to the outside. Platypuses and echidnas are the only living monotremes; they are believed to have descended from therapsids other than those that spawned the first true mammals and are sometimes called "specialized therapsids" as a consequence.
Multituberculate	"Many-bumped," a term referring to the rough surface of the teeth in the order Multituberculata of primitive rodentlike Mesozoic mammals. Multituberculates were extremely successful, disappearing only when the true rodents made the scene in the early Cenozoic.
Mutation	A heritable change in the genetic matter of an organism, resulting in a permanent change in the structure or behavior of the organism's descendants.
Myelencephalon	The portion of the hindbrain controlling the automatic functions of visceral organs in vertebrates.
Neopallium	The "new cloak," the expansion of the cerebral hemispheres characteristic of mammals and responsible for much of their mental acuity.
Neoteny	A condition in which infantile characteristics are retained into adulthood.
Notochord	In primitive chordates, a stiffening rod against which the muscles play in swimming; in advanced chordates the notochord disappears early in life and is replaced by a vertebral column.
Oligokyphus	"Few-humped," a small tritylodont therapsid with many mammallike characteristics that occupied an econiche comparable to those of many modern rodents.

Ophiacodont | "Snake-toothed," said of a suborder of primitive fish-eating pelycosaurs.

Ordovician | A period of the Paleozoic lasting from about 475 million to 425 million years ago and embracing the rise of the vertebrates in the world's waters.

Ornithosuchus | A swift bipedal thecodont archosaur closely related to the first dinosaurs. The name means "Bird-crocodile."

Paleopallium | The oldest part of the cerebrum, or forebrain, in vertebrates; primarily a center of smell.

Pangaea | "All-land," the giant protocontinent of which the modern landmasses are fragments. During the Permo-Triassic periods the southern part of Pangaea was richly inhabited by therapsids.

Pantothere | "All-beast," said of the subclass Pantotheria of primitive Mesozoic mammals from which metatheres and eutheres are descended.

Pariesaur | Any of various "wall-lizards," cotylosaurs that were among the first terrestrial vertebrates to experiment with herbivory.

Pelycosaur | "Bowl-lizard," any of the most primitive order (Pelycosauria) of synapsids.

Permian | The Paleozoic period lasting from 265 million to 230 million years ago and embracing the triumph of the pelycosaurs and the rise of the therapsids; the last period of the Paleozoic.

Photosynthesis | "Light-building," the process in which the sun's energy is trapped by green plants combining water and carbon dioxide to produce energy-rich sugar molecules; photosynthesis fuels the entire living system.

Phthinosuchus | "Waning-crocodile," one of the most primitive and most pelycosaurlike of therapsids.

Pinna | The external ear of mammals.

Placenta | In mammals an organ serving to transmit nutrients and wastes between the bloodstreams of the mother and the embryo.

Placoderm | "Plate-skinned," any of an extinct class of primitive fishlike armored vertebrates.

Polyphyodont | A condition in which teeth wear out and are replaced through life; contrast with *Diphyodont*.

Probainognathus | "Advanced-jaw," a genus of small therapsids lying close to the therapsid-mammal evolutionary boundary.

Probelesodon | "Before-dart-tooth," a genus of small therapsids lying close to the therapsid-mammal evolutionary boundary.

Prosencephalon | The forebrain of primitive vertebrates, originally a center of smell.

Reptile | Any of the class Reptilia ("Crawlers") of amniote vertebrates originally erected by Carl von Linné to include turtles, lizards, snakes, and crocodilians. Since von Linné's day, the class has been enlarged to include almost every extinct amniote not closely resembling living birds or mammals, whether or not such amniotes were "crawlers."

Rhomben-cephalon	The hindbrain of primitive vertebrates, a center for balance and the detection of vibration.
Scymnognathus	A genus of large primitive predatory therapsids with sharp, saberlike teeth.
Sphenacodont	"Wedge-toothed," any of the suborder Sphenacodontia of advanced predatory pelycosaurs that gave rise to therapsids.
Stapes	The "stirrup," a tiny bone that conducts sound to the cochlea in the inner ear.
Symmetrodont	"Symmetrical-teeth," any of an order Symmetrodonta of primitive Mesozoic mammals.
Synapsida	"Fused-arched Ones," the pelycosaurs and therapsids traditionally grouped together in a reptilian subclass that gave rise to the mammals.
Taxonomy	The art or science of classifying organisms in groups according to their descent and evolutionary relationships.
Telencephalon	In amniotes the front of the forebrain giving rise to the cerebral centers of smell and association.
Tetrapod	"Four-footed," any of the land-dwelling vertebrates.
Thecodont	"Socket-toothed," any of the primitive crocodilelike archosaurs that gave rise to true crocodiles and to dinosaurs and their relatives.
Therapsid	"Having-the-form-of-a-beast," any of the advanced synapsids of order Therapsida that gave rise to the mammals.
Theriodont	"Beast-toothed," any of the suborder Theriodonta of carnivorous therapsids.
Therocephalian	"Beast-headed," any of the infraorder Therocephalia of carnivorous therapsids.
Thrinaxodon	"Trident-tooth," a genus of small carnivorous cynodont therapsids of advanced mammallike form.
Titanosuchian	"Titanic-crocodile," any of an infraorder Titanosuchia of primitive, mainly carnivorous therapsids.
Triassic	The earliest period of the Mesozoic era, lasting from about 230 million to 185 million years ago. During the Triassic the synapsid line triumphed and was then replaced by the more efficient dinosaurian dynasty; mammals arose from small therapsids during this period.
Triconodont	"Triple-cone-teeth," any of an order Triconodonta of primitive Mesozoic mammals.
Tritylodont	"Triple-knob-teeth," any of a family Tritylodontidae of advanced rodentlike therapsids that survived well into the middle of the Mesozoic before being replaced by multituberculate mammals.
Turbinal	The scroll-like, paper-thin bones in the nasal cavities of mammals over which the odor-sensitive lining of the nose is distributed. Turbinals permit great enlargement of this lining, hence the olfactory acuity of mammals in general.

Tympanum   The eardrum in vertebrates.

Varanosaurus   "Monitor-lizard," a genus of primitive generalized pelycosaurs.

Vertebrate   Any of the subphylum Vertebrata of chordates having their nerve cords encircled by bony or cartilaginous rings called vertebrae.

Vibrissae   Whiskers, important sensory organs in nocturnal mammals, which spend a good deal of time feeling their way about.

BIBLIOGRAPHY

Other than this one, there are as yet (1980) no full-length books dealing specifically and totally with synapsid evolution. The interested reader is therefore referred to the publication departments of major museums such as the American Museum of Natural History, whose *Novitates* contain many articles of interest; similarly, the *Breviora* of the Museum of Comparative Zoology at Harvard, the *Postillae* of the Yale Peabody Museum, and other such museum publications throughout the world.

More general reading on vertebrate evolution may be had in the works of the late great Alfred Sherwood Romer (*Vertebrate Paleontology*, University of Chicago Press, 1968; *The Vertebrate Body*, Saunders, 5th ed., 1977) and in certain books by Edwin Colbert, notably *Wandering Lands and Animals* (Dutton, 1973), in which this gifted scientist traces the story of therapsid life on the protocontinent of Pangaea. Each of these books sports an extensive bibliography through which the reader may gain a broad look at the complex world of synapsid research.

INDEX

adaptive radiation, 1

amniote eggs, 17–18, 23

amniotes, 17–22. *See also* cotylosaurs; reptiles

amphibians, 14, 17, 18, 22
  temperature regulation in, 24

Amphioxus, 8, 9

anapsid ("archless") skulls, 28–30

anomodontia, 56. *See also* dinocephalians; dromasaurs

anteater, spiny, 118–19

anteosaurs, 69

Anteosaurus, 70

apsid ("arched") skulls, 28–30

archipallium, 112, 113

archosaurs, 89, 90. *See also* dinosaurs; thecodonts
  brain of, 103
  of Cretaceous period, 130, 131

arthropods, 6–8. *See also* insects
  transition to terrestrial existence, 14–15

articular bone, 106, 110

Australia, marsupials in, 128

balance and orientation, sense of, 104, 108–109

bats, 132

Bauria, 77

bauriamorphs, 76–79
  palate of, 76–78

beaver, 126

bipedalism, 93, 95

birds, 21
  in Paleocene epoch, 131–32

birth, live, 127

Blarina brevicauda, 128–29

body temperature. *See* temperature regulation

bones, growth patterns of, 124, 125

brain. *See also* cerebellum; cerebral cortex; cerebral hemispheres; neopallium
  of archosaurs, 103
  of first chordates, 9
  five-part, 102
  of mammals, 116, 132
  of therapsids, 101, 103, 116
  three-part, 101
  visual center of, 103, 110

breathing, 73
  in cynodonts and bauriamorphs, 77–78
  in gorgonopsians and therocephalians, 74, 76, 77

caseids, 42–45, 47

Cenozoic era, 131–32

cerebellum, 104, 106

cerebral cortex, 112. *See also* neopallium; paleopallium

cerebral hemispheres, 111–13

Chanaresuchus, 90

144

cheekbones of therapsids, 49, 50
Chelonia, 23. *See also* turtles
Chelydra, 103
Chiniquodontidae, 86–87, 93
chordates, 7–11
    encephalization of, 8–9
    jawless, 10–11
climate, early Permian, 41–44
cochlea, 109
cockroaches, 27
Coelophysis, 97
coelurosaurs, 95–97
color vision, 110–11, 129
coniferous plants, 67, 88
cotylosaurs ("stem-lizards"), 21–26
    adaptive radiation of, 22
    body armor of, 23–24
    increase in speed of, 26
    temperature regulation in, 24–26
Cretaceous period, 100, 128–30
crocodiles (crocodilians), 89, 92, 131
crossbow, history of the, as cultural ana-
        logue, 1–3, 70
cycads, 67
cynodonts, 76–87. *See also* Chiniquodon-
        tidae; cynognathids; gomphodonts;
        procynosuchids; thrinaxodonts
    jaws of, 78, 79
    molar teeth of, 79, 81, 84
    palate of, 76–78
    taxonomic dispute concerning, 79
    tooth replacement in, 81
    vibrissae (whiskers) in, 82–84
cynognathids, 85–86
Cynognathus, 86

dentary (tooth-bearing bone of the lower
        jaw). *See also* jaws
    of *Cynognathus*, 86
    of mammals, 107, 108
    of pelycosaurs, 106, 107
    of therapsids, 106, 110
Devonian period, 11, 15–17
diapsid ("two-arched") skulls, 29, 30
Dicynodon, 64
dicynodonts, 61–67
Didelphis, 127–28

Dimetrodon, 36–38, 40
dinocephalians, 56–61
dinosaurs, xii, 95–100
    of Cretaceous period, 130
    mammalian evolution and, 99, 100
diphyodonty, 120
docodonts, 125
dromasaurs, 67–68

eardrum (tympanum), 109–110
ears, evolution of, 106–110
echidnas, 87, 118–19
ectothermy, 20. *See also* temperature
        regulation
edaphosaurs, 31, 34–39
    predators of, 38–39
    temperature regulation in, 35, 38
eggs: amniote, 17–18, 23
    parental care of, 115–17
    retained within mother's body, 127
Emydops, 67
endothermy, 20, 84
Endothiodon, 66
Erythrosuchus, 93
Estemennosuchus, 58–60
euryapsid ("broad-arched") skulls, 29, 30
Eusthenopteron, 15
eutheres, 125, 127, 128
evolution, 1–5
    parallel, 39, 70
eyes, 110. *See also* vision
    of dromasaurs, 68
    of jawless chordates, 10
    pineal, 25–26
    transition to terrestrial living and, 13

fenestrae ("windows," or holes) in skulls,
        28–30
ferns, seed, 44, 63
fishes, sarcopterygian, 15–17
flowering plants, 88, 129
foramina in thrinaxodonts, 80, 81
forebrain, 101, 111

gait: of early mammals, 121, 124
    of *Erythrosuchus*, 93

of gorgonopsians, 71
of pelycosaurs, 33, 40, 45, 48
of therapsids, 45
Galesauridae, 80
garter snakes, 127
gill slits, 10–11
gizzards of thecodonts, 93
glaciers, 43, 44
Glossopteris, 44
gomphodonts, 85
gorgonopsians, 70–77. *See also* cynodonts
  breathing in, 74, 76, 77
  gait of, 71
  palate of, 76
Gracilisuchus, 95
gravitational force, transition to terrestrial
    living and, 13, 14

hair, 82–84. *See also* whiskers
Haptodon, 29, 34, 39
hearing, sense of, 104, 106–110, 115
hedgehogs, 120
hemichordates, 7, 8
herbivory (plant-eating), 47
  transition to, 34–35, 54, 56
hindbrain, 101–104
hominids, 133
Homo sapiens, x, 3–5
  adaptive radiation of, 133–34
  pineal body in, 26
humerus of pelycosaurs, 33–34
Hylonomus, 23
hyomandibular (bone), 109

ictidosaurs, 87
incus (bone), 106
inner ear, 106
inner space, 114–15, 132, 133
insect-eaters, 17–19, 23. *See also* cotylo-
    saurs; insectivores
insect-eating: adaptations facilitating, 26–
    28
  plant-eating compared to, 34–35
  by thecodonts, 92
insectivores (Insectivora), 128–29
insects, 14, 15, 17. *See also* arthropods
  increase in speed and agility of, 26
isopods, 14

jackal, black-eyed, 111
jaw joint: inner ear and, 106, 110
  of *Probainognathus*, 87
  as therapsid (reptilian)-mammalian
    boundary line, 21, 79, 86–87
jawless chordates, 10–11
jaws, xi, 11, 27, 80. *See also* cheekbones;
    dentary
  of cynodonts, 78, 79
  of dicynodonts, 61–63
  of pelycosaurs, 29, 31, 106
  of sphenacodonts, 39
  of thecodonts, 92–93
  of therapsids, 106–110
Jonkeriidae, 57
Jurassic period, 87, 100, 125

land-dwelling, transition to, 13–18
lateral-line system, 104
Linné, Carl von, 4, 20–21
live birth, 127
lizards, 24
  pineal eye in, 25
lumbar vertebrae, 121, 124
lungs, 12, 15. *See also* breathing
Lycosuchus, 71, 74–75
Lystrosaurus, 63–65

malleus (bone), 106
mammals, ix–xi
  bone-growth patterns in, 124, 125
  boundary line between therapsids
    (reptiles) and, 21, 79, 86–88
    brain of, 116, 132
  cerebral hemisphere of, 113
  dentary of, 107, 108
  dinosaurs' role in evolution of, 99, 100
  first, 120–31
  hearing sense of, 106–110
  neopallium of, 112, 116
  in Paleocene epoch, 131–32
  skeleton of, 122–23
  skull of, 105
  teeth of, 100, 120, 126
  tree-dwelling, 129
  vision in, 110–11
marsupials, 128. *See also* metatheres
masseter (chewing) muscle, 86, 106

Massetognathus, 84
maternal feeding of the young, 117–18, 120
Mesozoic era, 95–101, 128. *See also* Cretaceous period
   dimming of sunlight at the end of, 131
metatheres, 125, 127–28
milk glands, 127
molar teeth, 106
   of bauriamorphs, 78
   of cynodonts, 79, 81, 84
   of *Massetognathus*, 84
   of thrinaxodonts, 79–81
Mollusca, 6–8
monotremes, 87–88, 118–19
Morganucodon, 121
multituberculates, 120, 125, 126
myriopods, 15

neopallium, 112–16, 132, 133
neoteny, 8
nerve cord, in early chordates, 7–10
nipples, 127
nostrils, 112
notochord, 7, 8, 10

Oligokyphus, 88
Ophiacodontia, 31
opossums, 127–28
Ordovician period, 6–10
Ornithorhynchus, 118–19
Ornithosuchidae, 93

palates, 76–78, 80
Paleocene epoch, 131–32
paleopallium, 112, 113
Pangaea, 40–44
pantotheres, 125, 127
parallel evolution, 39, 70
parental care of eggs, 115–17
parental feeding of young, 117–18, 120
Pariesauridae, 47
pariesaurs, 55
pelycosaurs, xi, xiii, 29–48, 50. *See also*
   caseids; edaphosaurs; Ophiacodontia;
   sphenacodonts
   climate changes in the Lower Permian
     and, 44
   dentary of, 106, 107

gait of, 33, 40, 45, 48
"half-pushup" posture of, 33, 38–39
humerus of, 33–34
jaws and teeth of, 29, 31, 106
skull of, 105
way of life of, 40–41
Pennsylvanian period, 26
   early, 21, 24
Permian period: early, 31, 40–47
   middle, 50–54, 61, 69
Phthinosuchus, 48
phytosaurs, 89
pineal body, 10, 26
pineal eye, 25–26
pinnae, 110, 111
placenta, 127
Placodermi, 11
plankton, 130
planning, 114
plant-eating (herbivory), 47
   transition to, 34–35, 54, 56
plants: coniferous, 67, 88
   of Cretaceous period, 129–31
   early Permian, 43–44
   flowering, 88, 129
   transition to land-dwelling, 14
   of Triassic period, 88
platypuses, 87, 118–19
poison glands in insectivores, 128–29
poisonous bite of therocephalians, 71
posture: of pelycosaurs, 33, 38–39
   of pseudosuchia, 93
   of titanosuchians, 52
pressure–change–adaptive radiation–pressure, cycle of, 1–3, 133–34
primates, 129, 132
Probainognathus, 86–87
Probelesodon, 83
procynosuchids, 80
pseudosuchia, 93

quadrate (bone), 106, 110

Reichert, C., 106
reproductive process. *See also* eggs
   transition to terrestrial living and, 14,
     17–18

reptiles, 103. *See also* cotylosaurs; therapsids
    definition of, 20–21
rhynchosaurs, 55
ribs of therapsids, 121

"sails": of edaphosaurs, 35–38
    of sphenacodonts, 39–40
sarcopterygian fishes, 15–17
scorpions, 14–15
Scymnognathus, 72–73
seed ferns, 44, 63
selective pressure, 1. *See also* pressure–change–adaptive radiation–pressure, cycle of
seymouriamorph, 19
shrews, 120, 128–29
sight. *See* vision
Simpson, George Gaylord, 100
skin: temperature regulation and, 82
    of therapsids, 118
skin secretions, feeding of young with, 118, 127
skulls, 27
    anapsid ("archless"), 28–30
    apsid ("arched"), 28–30
    of bauriamorphs, 77
    diapsid ("two-arched"), 29, 30
    of dicynodonts, 61, 63–67
    of *Dimetrodon*, 38
    of dinocephalians, 56
    of *Emydops*, 67
    of *Endothiodon*, 66
    euryapsid ("broad-arched"), 29, 30
    of *Lystrosaurus*, 64–65
    of snapping turtle, 104
    synapsid ("fused arch"), 29, 30
    temporal openings for jaw muscles in, 105
    of thecodonts, 92
    of *Venyukovia*, 60
smell, sense of, 9, 101, 111–14
snakes, garter, 127
snapping turtles, 103, 104, 116
South America, metatheres in, 128
South Pole, 43
sphenacodonts, 36, 39–41, 47
    teeth and jaws of, 39

temperature regulation in, 39–40
spinal column, 121
stapes, 109
stereoscopic vision, 129
suckling, 118, 127
sunlight, late-Mesozoic dimming of, 131
swamp-dwellers, 15–17, 22
symmetrodonts, 120
Synapsida, xiii, 26, 29. *See also* pelycosaurs; therapsids

Tachyglossus, 118–19
tails of dromasaurs, 67
Tapinocephalus, 61
taste, sense of, 113
taxonomy, 20, 21
teeth. *See also* dentary
    canine, 31, 38, 39
    deciduous ("milk"), 120
    of dinocephalians, 56, 60
    mammalian, 100, 120, 126
    molar, *see* molar teeth
    of pelycosaurs, 29, 31
    replacement of, 81, 117
    of sphenacodonts, 39
    of therapsids, 117, 126
    of therocephalians, 71
    of titanosuchians, 51, 57
    of tritylodonts, 126
temperature changes on land, 13–14
temperature regulation. *See also* endothermy
    in cotylosaurs, 24
    in cynodonts, 82–84
    in edaphosaurs, 35, 38
    in gorgonopsians and therocephalians, 75, 77
    pineal eye and, 25–26
    in reptiles, 20
    skin and, 82
    in sphenacodonts, 39–40
    in titanosuchians, 52
    in turtles, 25
temporal muscles, 106–107
tenrecs, 120
terrestrial living, transition to, 13–18
tetrapods, 12
Thamnophis, 127

148

thecodonts, 89–95
  bipedalism among, 93, 95
  skulls of, 92
  therapsids and, 92–95
therapsids, xi, 34, 49–57, 99–120. *See also*
  Anomodontia; bauriamorphs; cynodonts; gorgonopsians; therocephalians;
  titanosuchians; *and other specific ther-*
  apsids
  adaptive radiation of, 52–53
  bone-growth patterns in, 124, 125
  brain of, 101, 103, 116
  cerebral hemispheres of, 112, 113
  cheekbones of, 49, 50
  dentary of, 106, 110
  econiche of, 101
  evolutionary tree of, 47
  gait of, 45
  jaws of, 106–110
  odor sensation in, 111–14
  parental care of young among, 115–17
  parental feeding of young among, 117–
  118, 120
  plant-eating, 54–57
  skeleton of, 122–23
  skull of, 105
  teeth of, 117, 126
  thecodonts and, 92–95
Theria, 125
Theriodontia, 54
therocephalians, 70–77. *See also* bauriamorphs
  breathing in, 74, 76, 77
  palate of, 76
Thrinaxodon, 79–82
thrinaxodonts, 79–82

tooth replacement in, 81
titanosuchians, 50–54, 57, 69–70
  posture of, 52
tree shrews, 129
Triassic period: plant life of, 88
  Upper, 41
triconodonts, 120
tritylodonts, 87, 88, 125, 126
tundra, 43–44
Tupaia, 129
turbinal (bones), 112, 114
turtles, 131
  armor of, 23–24
  snapping, 103, 104, 116
  temperature control in, 25
tusks of *Lystrosaurus*, 63
tympanum, 109–110

Ulemosaurus, 56

Varanosaurus, 29, 32–33
vegetarianism. *See* plant-eating
vegetation. *See* plants
Venyukovia, 60
vertebral column, 10
vertebrates, 6
vibrissae. *See* whiskers
vision. *See also* eyes
  brain and, 103, 110
  color, 110–11, 129
  stereoscopic, 129

walking. *See also* bipedalism; gait
  of titanosuchians, 52
whiskers (vibrissae): in cynodonts, 82–84
  in thrinaxodonts, 80–82

DATE DUE

JUL 1 1 1983			
MAR 2 3 2002			
APR 2 1 2002			